E. C. Stakman, Statesman of Science

Elvin Charles Stakman, 1885–1979. (Courtesy, Cosmos Club, Washington, DC)

E. C. Stakman, Statesman of Science

C. M. Christensen

Regent's Professor Emeritus
University of Minnesota

Published by
The American Phytopathological Society
St. Paul, Minnesota

Financial support for this book was given by the Rockefeller Foundation.

Library of Congress Catalog Card Number: 84-70114
International Standard Book Number: 0-89054-056-X

Printed in the United States of America

The American Phytopathological Society
3340 Pilot Knob Road
St. Paul, Minnesota 55121, USA

Contents

Foreword

Dr. Elvin Charles Stakman was a brilliant scientist, formidable scholar, prodigious educator, and a compassionate humanitarian. Without doubt, his many monumental discoveries and contributions to the study of plant diseases and the control of crop losses qualify him as one of the founders of modern plant pathology.

Much of his research was devoted to studies of crop losses caused by plant pathogens, especially by rust and smut fungi and organisms that cause other foliar diseases and root rots of wheat, oats, and maize. His research interests transcended diseases of cereal crops, however, and ranged from diseases of potato to those of the Brazilian rubber tree.

His fundamental pioneering studies on the nature of genetic variation of the stem rust pathogen, *Puccinia graminis* f. sp. *tritici*, established that this pathogen was made up of many physiologic races, which differed in pathogenicity. This discovery explained why new commercial wheat cultivars continually lost resistance to rust as shifts occurred in the races of the rust population. The studies led to the establishment of a sound, scientific basis for effective wheat breeding programs in the United States and in other countries. In fact, it ushered in a new era in the improvement of wheat and other cereals. It also established a scientific model upon which to organize breeding programs to develop disease resistance in many other crops.

Stakman also studied the epidemiology of wheat stem rust. This led to an understanding of the factors that interacted to cause rust epidemics on wheat both locally and regionally. His studies established that vast epidemics, such as those that devastated wheat on the Great Plains in 1904, 1916, and 1954, were brought about by urediospores that survived the mild winters of southern Texas and northern Mexico and were blown northward into the winter and spring wheat regions of the United States and Canada.

Without doubt, Dr. Stakman was one of the greatest teacher-educators of this century and perhaps all centuries. His effectiveness was attributable, in large part, to his broad interests and vast knowledge combined with an incredible memory for details. He had the unique capacity to provoke, excite, and stimulate most students to reach for a star that they thought was beyond their intellectual limits. A surprising number of his students at least got some "stardust" on their hands and became outstanding leaders in their chosen fields in many countries of the world. But Stakman, the warm, concerned human that he was, also recognized his students' limitations and guided them into positions where they would fit and be happy and in which their individual strengths and skills would be most gainfully used.

His graduate seminars generated provocative, stimulating discussions and transmitted information concerning many disciplines and spheres of human endeavor. Equally edifying were the informal discussions that occurred between 10 and 11 p.m. as Stakman, with pipe in hand, made his biweekly strolls through the graduate research laboratories, presumably to learn about new research discoveries, but more likely to see who was still at work.

From a lecture platform, Stakman was a highly effective, eloquent advocate for agricultural science and especially for plant pathology. During my last month of undergraduate study as a forestry student at the University of Minnesota, I attended a Sigma Xi lecture presented by Dr. Stakman entitled "Plant Diseases Are Shifty Enemies." In one hour, he converted me from a forester to a forest pathologist, and a year and a half later he converted me into a general plant pathologist.

Stakman was involved in the establishment of the Cooperative Mexican Government-Rockefeller Foundation Agricultural Program in 1943. This technical assistance program, which trained young Mexican scientists and conducted research on Mexico's principal food crops, was so successful that it led to the development of a worldwide network of agricultural research centers, including the International Maize and Wheat Improvement Center (CIMMYT) in Mexico, where I have spent many years applying principles established by Stakman or based on his work.

In 1980, while I was revisiting the University of Minnesota, I sat one day in the Plant Pathology Department drinking coffee with Drs. Carl Eide and Dave French and reminiscing about the state of plant pathology and education in general in the late 1920s and 1930s. I am certain that most of the graduate students present were certain that most of these recollections were pure fabrications by a few living fossils. Nevertheless, as I have traveled and worked in many countries over nearly 40 years, I have repeatedly been approached by scientists who have introduced themselves as having studied under Stakman, J. J. Christensen, Carl Eide, Helen Hart, Matt Moore, Clyde Christensen, Tom King, and Milt Kernkamp in the old "Tottering Tower" or the "New" Plant Pathology Building. There is no question but what the Department of Plant Pathology, especially under the stimulation and prodding of "Stak," cast its shadow of scientific influence and intellect around the world. Reflecting on Dr. Stakman's tremendous influence, Carl, Dave, and I decided to ask Clyde Christensen, one of Stak's colleagues and a longtime friend, to write a biographical sketch of our great mentor and friend. I am pleased that this book is the result.

Norman E. Borlaug

College Station, TX
January 10, 1984

Preface

I first met E. C. Stakman in 1928 when, after my junior year in the Forestry School at the University of Minnesota, I worked during the summer as a field and laboratory assistant in the Department of Plant Pathology. During my senior year, I took the course in forest pathology that all forestry students were required to take and that, by some oversight, I had neglected to take in my sophomore year, fortunately, as it turned out. I liked the course, and, as I did very well in it, Stakman thought I might be a suitable candidate for graduate work in plant pathology. The position of Instructor in Forest Pathology was about to become open, and Stakman offered it to me. I accepted with alacrity. If anyone in the administration had any qualms or objections about appointing a young man with only a B.S. degree to a full-time instructorship, Stakman must have quieted them. In the fall, I began my career in plant pathology as instructor, in charge of teaching, research, and extension in forest and tree pathology. Thanks to Stakman, I was now, with one undergraduate course in forest pathology to my credit, a professional plant pathologist.

During the next 15 years, except for the year I spent as an exchange student in Germany, an arrangement made by Dr. Stakman, my major work was in forest pathology. In the mid 1940s, the department was asked to undertake work on several problems involving fungi in cereal grains and their products, and Dr. Stakman suggested that I undertake responsibility for carrying on this work. One of the problems dealt with loss of quality caused by fungi in stored grains. This turned out to be important, not only in the United States, but wherever grains were stored, including Mexico.

Through Dr. Stakman's recommendation, I was invited by the Rockefeller Foundation in 1959 to visit Mexico to get work under way on grain storage problems there. This first visit led to several others and to work with Mexican graduate students. The work on fungi in stored grains continued to occupy most of my official and much of my unofficial time until my retirement in 1974, and even beyond, until I left St. Paul for Arizona in 1980.

Obviously, Stakman had a tremendous influence on my career, on my professional development—and, through innumerable discussions over nearly 50 years, on my intellectual development and outlook in general. He was, of course, my major advisor in my graduate work, but he was much more than that—preceptor, counselor, close friend, and associate, a person with whom I could share problems and successes and failures, and one whose door and mind, both at work and at home, were always open to me.

In the fall of 1980, a year and a half after Stakman's death, Dr. Norman Borlaug asked me to write a biographical sketch of Stakman. Dr. J. George Harrar, then president emeritus of the Rockefeller Foundation, concurred with Borlaug's suggestion. Dr. D. W. French, head of the Department of Plant

Pathology at the University of Minnesota, sent to my home a wealth of documents and records dating back to the founding of the Department in 1907.

In 1970, Stakman had dictated an oral history of about 1,685 pages at the Rockefeller Foundation offices in New York, and a copy of this was made available to me. It furnished not only interesting background material about his youth and college days but also detailed accounts of some of his later activities. The present work quotes from it liberally and also from Stakman's many research papers and his later general writings, some of them classics, on science and society and on work and problems in agriculture. Thus, much of what follows is in Stakman's own words.

In the later years of his long and productive life, Stakman became a spokesman for the role of science, and particularly agricultural science, in the solution of some of the basic problems that beset mankind. In recognition of this, the Cosmos Club Award in 1964 was entitled, "Elvin Charles Stakman: Distinguished Biologist, Great Teacher, and Statesman of Science." Like so many of his fellow agricultural scientists who had extensive first-hand experience with the people and problems of the developing countries, Stakman was very much aware of the food–resources–population dilemma that confronted so much of the world then, and continues to confront it today.

Stakman obviously had developed an international outlook long before he became associated with the Rockefeller Foundation program in Mexico—it was this that helped make his contributions valuable there—and this outlook grew stronger and more robust and explicit with time. He was an effective proponent and promoter of international cooperation on problems common to different countries, and he emphasized again and again in his speeches and writings throughout his last 30 years the need for international effort and support to solve these problems. Stakman's genius was not just of his time and place, but in any time would have illuminated any society that respected love of learning and the urge to put knowledge to use for the benefit of mankind.

It is a pleasure to here express my gratitude to a number of friends who have been instrumental in making this book possible: To Dr. Norman E. Borlaug, International Corn and Wheat Improvement Center, Mexico, who first suggested that the work be undertaken and who gave valued help in the revision of the preliminary drafts, especially of those portions dealing with Stakman's involvement in the work in Mexico; to Drs. D. W. French, C. J. Eide, and T. Kommedahl of the Department of Plant Pathology, University of Minnesota, for materials from the Department and from Stakman's personal files, and for criticism of the manuscript during its preparation; and to Dr. John A. Pino, Marjorie J. Schad, and the late Dr. J. G. Harrar of the Rockefeller Foundation, for making available the oral history and other records, published and unpublished, and for their encouragement and support. To all of them, my most sincere thanks.

C. M. Christensen

Sun City West, AZ
January 30, 1984

PART I

The Early Years

Stakman, during the time he lived in Brownton.

CHAPTER ONE

Boyhood

Elvin Charles Stakman was born May 17, 1885, to parents Frederick and Emelie Eberhardt Stakman. According to his own account, dictated in 1970 in an oral history, "I was born at Ahnapee, Wisconsin, which was the eastern terminus of the Ahnapee, Green Bay and Western Railroad, which was about 28 miles long and which was named after the Ahnapee band of Indians." His birthplace was a farm overlooking Lake Michigan. The town was later named Algoma, and this is listed as his official birthplace.

Elvin was the youngest of four children; the others were Arthur, born in 1878; Lawrence, born in 1880; and Edna, born in 1883. Arthur was characterized by Elvin as "a fine, strong man." He played semiprofessional baseball, attended a business college in Green Bay, from which he graduated with honors, and became a purchasing agent. Lawrence, the next oldest, had "elements of mechanical genius" and even as a youth was able to construct steam-powered machines that actually ran. He became a "traveling engineer" on the railroad. Of his sister, Stakman said, "I owe her a very great debt of gratitude for having helped me when I needed help." She started teaching school when she was 17 and "had the reputation of being a very exceptional primary teacher." Nowhere in the oral history is the father mentioned.

The family moved to Brownton, Minnesota, while Stakman was still "a babe in arms." Brownton, then a town of 350 inhabitants, is in McLeod County, about 75 miles west of Minneapolis and St. Paul and at that time was just where the forests of the "Big Woods" abruptly gave way to the prairie. In 1971, Stakman wrote to the editor of the *Brownton Bulletin*, the town paper:[1]

Hundreds of times and in many places I have told others how lucky I was to have lived during my boyhood and youth in the natural and human environment of Brownton. The native woods and virgin prairies, the lakes and sloughs, the ponds and creeks, the meadows and pastures, and the fields and farms were always an invitation to enjoyment and to learning. And the human environment was equally interesting and even more significant in many respects. It comprised Yankees and other old-line Americans, English, Irish and Scotch, Germans and Danes, Bohemians and Poles. But all were Americans, thanks to the general desire for opportunity and for responsible self-government and to several unifying influences, especially the school, the town meeting, Christmas, Memorial Day, and the Fourth of July. I learned something from all of these various people, and many of them taught me priceless lessons by word or by deed. Unfortunately it is now too late to thank most of them, for they are gone. But my gratitude to them

3

and to those still living will always live within me.

I am proud that I am still considered a "Brownton native," for in grateful memory I still live there much of the time.

Even granting some nostalgic hyperbole in this, it still is a fine tribute from a big man to a small town, and it certainly differs greatly from the evaluation of another small Minnesota town of a slightly later era, by Sinclair Lewis in *Main Street*.

Brownton had been founded a few years before the Stakmans moved there by Captain Alonzo Leighton Brown and was named after his brother Charles, who had been killed in the battle of Shiloh. Many veterans of the Grand Army of the Republic lived there, and on winter evenings Stakman listened to their reminiscences around the potbellied stove in the village store, where he worked as general factotum and clerk. Even as a youth, he attended town meetings, listening with interest to the discussions, and he absorbed lasting impressions of politics and politicians and of the democratic process. As a boy of seven, he shared the general disappointment of the community when the Democratic candidate for the presidency, Grover Cleveland, defeated the Republican candidate, Benjamin Harrison. Brownton was Republican.

Brownton was a farming community, and Stakman early became aware of the vital importance of agriculture in the life of the people. In the oral history,[2] he describes some aspects of farm life then:

> I think insofar as the economic status was concerned, much of the farming was subsistence farming. There were some people that got rich buying and selling land, but the primary objective on the farm was to make a good living, and the farms, many of them, were self-sufficient. They had their own blacksmith shops, little ones, but a blacksmith shop, a carpenters shop, a smokehouse where they cured meat for the winter and smoked it. They did their own slaughtering of farm animals and many of them had their own sheep and carded the wool, and knitting was a very common occupation for women. They were seldom idle, clicking their knitting needles all of the time, and they knitted socks and that sort of thing, although at that time the buffalo robe was the winter garb. Buffalo caps, buffalo coats, buffalo boots and all that sort of thing. You couldn't get cold inside of those and they were dirt cheap. I think I remember when a buffalo coat cost about eight dollars, and I think now . . . . the last one that I heard of anybody having cost eight hundred dollars instead of eight. So times have changed.

Social life and recreation were important too:

> The people lived so close to each other that everybody knew everybody else. They gossiped, but this was not unfriendly gossip, mostly. They had to have something to talk about. They were interested in their neighbors. Their neighbors knew what was going on. Some of the New England Yankees might "deacon" a barrel of crab apples or something like that, but there was humor in it. It was a little bit like the David Harum sort of thing. It was a game. They'd out-slick each other if they could. But mostly it was good-natured. Cheating, dishonesty of any kind, was anathema, because it wasn't neighborly; neighborliness and the need for cooperation in communities like that, a pioneer community, was awfully important. And so children, everybody, learned something about human subsistence and about the need

for good human relations. I think that had a very strong influence on many people.

Insofar as the recreation was concerned, this again I think is tremendously important. Even we youngsters had to make our own entertainment. They didn't organize baseball teams for us, and they didn't organize any teams for us. We organized them ourselves and we got experience in doing it. We played for fun. We played hard, but we played for fun. I think at one time in this little town when there weren't any more than 400 people in the town, there were six baseball teams, starting with little toddlers and going up to the so-called first team—first team, second team and so on down to the sixth team. All of the time that I can remember there were at least five teams, and even when I was on the fifth team I remember we organized our own team, elected our own captain, got one of the older boys to coach us if we could get him to help a little, but this really was self-entertainment. We would hire a livery rig if we needed to or we'd get somebody to donate a rig and we'd go to New Auburn, Stewart, and various towns around and play our games of baseball.

Hunting, of course, was important, and fishing. There were a lot of lakes around there and people hunted and fished for meat. And there were lots of prairie chickens; there were wild geese, wild ducks, and rabbits and many other kinds of game. I don't think that people really considered it was a sport to get a fish on the hook and play him for 10 minutes. It was partly the satisfaction of getting a different kind of meat and feeling that you still had good aim if you needed to. And so there was a lot of hunting and a lot of picnicking, school picnics, church picnics, and town picnics, and there was some sailing and boating on the lakes. In the wintertime there were skating and sleigh riding and candy pulls and that sort of thing. It was a very intimate kind of social life, and yet people didn't intrude on each other's privacy. If they wanted to ask questions, they usually asked them with a considerable amount of discretion.

As to schools:

The standards I think were good. The educational standards were high. Everybody was supposed to get an education, and it was a very common saying among parents, "I want you to have a better education than I had, because I've missed not having one, but I didn't have the opportunity you have. Now, you get an education."

My mother was a good school teacher. She wasn't a professional school teacher, but she certainly tried to help us youngsters learn. I remember we had regular quiz sessions. We would say, "Ask me another, ask me another," and if we didn't have the answer she'd try to get us to look for it. Also I remember memorizing poems for declamation. She said I started reciting when I was about three-and-a-half years old. I don't know, I'm not sure. I remember doing it when I was four. . . . I started visiting school when I was scarcely four years old. "Aunt" Polly Baker was the teacher. The primary classes were on the first floor of the Odd Fellows Hall, and there was a two-room schoolhouse nestled in an oak grove half a block away, and that was a fascinating place. It was a very attractive building. I'll never forget it.

But this was a wonderful experience, to go to school to "Aunt" Polly Baker. I don't know how much education "Aunt" Polly had; I'm sure she never took any pedagogy because they didn't teach much pedagogy in those days. But she did what every good teacher should do. She took an interest in the kids. She tried to get them to learn in every way. Her discipline was for the most part kindly, and she'd give everybody a chance.

I remember she asked me one time—I wasn't going to school officially yet but was permitted to visit—where my heart was. I told her it was in the left side of my brain. I corrected myself very quickly but I never lived that one down. I spoke too fast. And she said, "Don't speak faster than you can think." This was a good lesson, of course.

I also remember another lesson in caution. There was a boy by the name of Jim Ebling and he was probably about 10 when I was four, and at recess we'd see how far we could run and get back on time because there was a little hand bell that she used to ring when recess was over. Well, we were quite a way away, and I think I told you that going out and playing in the rain hadn't lengthened my legs particularly [he had been told that if he went out in the rain and got wet, his legs would get longer and he would be able to run faster] and hadn't increased my speed, so I couldn't keep up with some of the bigger boys. I was running along behind them, and a beautiful big piece of chocolate dropped out of the blouse of this older boy, and I was running fast enough so that I ran past it, and then I ran back to pick it up, and I got back to school a little bit late. But I went to Jim's seat and I said to him, "Here, Jim, you lost this on the way down," and I saw "Aunt" Polly look at him with a scowl on her face and she said, "Well, now, I'll take that." After school, for some reason, it was a mystery to me, Jim hit me on the side of the head and knocked me into a patch of weeds. I think I rolled over six times. And I discovered later that the "chocolate" was a big plug of tobacco, which he had swiped from his father's store. Apparently he was not supposed to chew tobacco in school.

Well, okay, when "Aunt" Polly explained to me what had happened, I decided to mind my words thereafter.

Even in grade school he was evaluating his teachers, and not all of them came through with high marks:

There were some teachers who were unreasonable. Some were afraid to say that they didn't know, and you began to detect that, and you kept saying to yourself over and over again, "If I ever teach school, I'm not going to teach like that teacher teaches."

Probably no teacher or teaching, however bad, could have dulled or stifled young Stakman's urge for learning and quest for knowledge. If the materials for learning were there, he would have learned, regardless. Even in grade school he was setting some pretty high standards for his instructors, and it is no wonder that some of them did not quite measure up in all respects. To continue:

Those were the days when "licking and learning" still went together, and there were certain punishments for certain misdemeanors. I remember they used to use a big maple brass-edged yardstick to spank your hand with until some of us got in the habit of putting out our hand and then jerking it away just as the ruler came down, and it came down on the desk, and then there was trouble.

I want to say one other thing about this dogmatism that many of the teachers had—"Well, I'm ashamed of you because you don't know that." Instead of telling you how to find out and encouraging independence, when you'd ask them, "Why is this so?" they would simply say, "Because the book says so."

Now there were others who went along with you and tried to help you learn. I remember one time that the teacher was trying to bring out the meaning of the word "oval," and I wasn't sure what she was driving at, so she said, "What shape is an egg?" I said, "An egg is round," and she kept on trying to get me to say that it was oval. The last eggs I'd seen were round. So she said to one of the boys, "You go on over to 'Aunt' Polly Baker's and get an egg." He brought it over. She said, "Huh, what shape is that?" He said, "Oval," and she said, "Uh-hum, I thought *you* might have known that, Elvin." I said, "Well, listen, teacher, that egg *is* oval, but you didn't ask me what shape a hen's egg was, you asked me what shape is an egg. The last eggs that I've seen were mud turtle's eggs, and I've seen a lot of fish eggs, and they're round, and there are certain birds that lay round eggs. You didn't say anything about a hen's egg. Why should I assume that you were talking about a hen's egg?"

Well, I think that for about three weeks I was in the doghouse, at home and at school. Now I think that that was an educational crime. I think that they should have said, "Okay, maybe you were trying to be a little smartalecky, but..."

Stakman does not say at what age this spirited exchange about round and oval eggs took place, but it can hardly have been later than the third or fourth grade. His mental sharpness and love of argument were already fairly well developed. Had the teacher specified "hen's egg," he could have had her there, too, since "hen" is applied to some kinds of fish and to some kinds of crustaceans that lay round eggs, as well as to the domestic fowl, which lays oval eggs. I would also question the authenticity of his injured innocence. Children, especially smart ones, are very good practical and practicing psychologists and delight in baiting and bedeviling their elders and betters. Stakman was a perceptive boy and must have known that his teacher was not likely to have seen or to be familiar with turtle or fish eggs, and that she was referring to the egg of the common barnyard fowl, the egg that she and others handled and ate. He was being obstructive and contentious, and if he was put down afterwards, this was no more than simple justice, not an educational crime. Even in grade school, Stakman must at times have been a rather formidable opponent in confrontations of this sort and, at times, somewhat of a trial to a tired and harassed teacher. Throughout his long life, argument for the sake of argument was one of his pleasures, and frequently when he had an opponent on the ropes he would gradually maneuver and shift around until he was on the opposite side of the argument and then beat his opponent on that one, too.

Like many another boy, Stakman herded cattle as a youngster:

I started herding cattle when I was just under seven, and I herded cattle for eight years—that is, during the growing season—because a boy could do almost as good a job as a man, and so my brother and I herded the town herd for five years together, and then there was another boy and I that had the town herd for one year, and then I herded cattle on the Dwinell Ranch all or part of the additional two years.

The cattle involved were most likely the somewhat scrubby milk cows of the time, and the herding consisted mostly of seeing that none of the cattle wandered off as they grazed. It gave the young herder plenty of time to get acquainted with the prairie flora and fauna, especially the flowers and birds. Stakman took full advantage of this, since he mentions many of the prairie and woodland flowers and birds by name. If he recognized as many of these as he appears to have, he was an observant boy, as there is every reason to suppose he was. Many of the pioneers, who lived close to nature, were sensitive to the beauties of wild plants and animals, and in every community someone was likely to be familiar with the common names of the wild plants. Familiarity with the wild animals that abounded there came naturally.

Stakman very early realized that the virgin forests and virgin prairies around Brownton were precious and irreplaceable but headed for extinction. At age 10, he wrote an essay on "What the Axe and Fire Did to the Woods" and another on "What the Plow and the Ditcher Did to the Prairie." He does not say whether the choice of these subjects was his own or whether they had been suggested by a teacher. It would be interesting to read those essays now and to know the sources of his information. Whatever thoughts and views were expressed, they were most likely his own and probably were formed by thoughtful consideration of all available information, including that which he gained from his own personal observation and experience.

Like all boys, Stakman had odd jobs and chores to do, too. One of these was sawing up cordwood for home cooking and heating:

> You cut the four-foot pieces into three pieces with an old bucksaw. You had a sawbuck, two X's with a crossbar. You put the wood in that, and you'd saw it with your bucksaw. Sawing wood was a tedious sort of job, but you knew how the different kinds of wood sawed. You'd get ironwood and you'd say, "Well, I'm not going to saw for a while." Your mother would say, "You'd better get that wood sawed." "Aw, this ironwood, it turned the teeth of the saw so I've got to file the saw." You had a little triangular file and you filed the saw to sharpen it.

Actually, filing a saw, even a bucksaw, to sharpen it requires considerably more expertise than a schoolboy, even an unusually bright one, was likely to have. Stakman was not handy with tools of any kind—he just was not mechanically adept; he might have been able to give a detailed and plausible account of how a saw was to be sharpened, but I doubt if he ever filed a bucksaw blade to sharpen it. A small matter. To continue:

> You learned about wood when you were sawing it, and then when you were splitting it you also learned a good deal about it, because you learned whether it was cross grained and tough or whether it was nice uniform texture. You learned the color. You learned the smell of it, if it had an odor of any kind, and you learned about its qualities. You picked out good whittling wood. Whittling was quite an occupation in those days. You may remember that General Grant used to sit and whittle twigs when he was pondering the tactics in the battles, many of the battles, of the Civil War. He was a great whittler. Whittling was a very common—well, sedative, I suppose, for people, and it was sort of nice. If you got a good piece of butternut or a good piece of red cedar or something like that, you always

had a sharp knife and you'd just whittle that and polish it, and so you learned a lot about wood.

Well, finally I remember I decided it would be nice to whittle wood and at the same time to learn more about the wood, so I started whittling books out of wood, and [I'd] print the name of the wood on the back, so that I had a library of woods. And I taught wood technology to the students in the College of Forestry[3] many years later, without ever having had a course in wood technology. I had learned something by that time about wood anatomy and histology, and I pieced that together with the knowledge that I got when I was a boy, of woods, in a practical way. They said it was a good course. I'm not sure that it was, but at any rate I had a lot of fun teaching it. But this I never forgot—the value of firsthand knowledge.

As a boy in Brownton he even accumulated some knowledge of microbiology:

We didn't have very good refrigeration in those days, and so you'd see food spoil, and you'd wonder what spoiled it. You'd see the molds on food and you'd see slimy growth and that sort of thing. Well, that made me interested in the mechanism of food spoilage. I used to wonder about it as a youngster. There wasn't anybody that ever told me very much about bacteria, although we did learn something about them in school, and in "current events" we learned about Pasteur and about Robert Koch, because in those days the germ theory of disease was just becoming firmly established.

But all of these things made you realize that natural history, that biology, was not only tremendously interesting but could also be extremely useful. So when I was asked to teach a course in household bacteriology in the university to about 90 home economics students, I already knew quite a little about food spoilage. In the meantime I'd learned some bacteriology and mycology. Well, you can imagine what this was—a plant pathologist in a predominantly man's school teaching 90 home economics girls who'd never had a course in science. They almost mutinied, and complained to the dean that I was going to flunk them all, that nobody'd ever pass; but actually at the end of the course they had learned a good deal. Among other things, they'd learned something about working intellectually.

In 1979, Esther Tolaas, who in 1913 had been a student in the course in household bacteriology taught by Stakman, recalled:[4]

Stak was a stern taskmaster. I will always remember his keeping us in the laboratory until after dark, cleaning the equipment. Many of the girls were in tears, especially some who commuted by street car between the campus and their homes in St. Paul or Minneapolis and had to spend 2 or 3 hours each day riding the rails. In the winter I left home before dawn and got home long after dark.

Stakman also acquired a considerable knowledge of the German language as a boy in Brownton. There were many first-generation German-Americans in the community, and some of their children were bilingual, so he picked up some of the language from them. He learned it also from "old Mr. Schimpelpfennig," the school janitor, and in the village store where he worked he had to be able to speak and understand some German to deal with the German-speaking farm wives who

9

came in. For two summers he was a paying pupil in the Lutheran Church school, which he attended specifically to learn German. By the time he went to the university, he was at least fairly well versed in spoken and written German. When he graduated from the university, German was a usable second language for him, and he never lost his love of German literature. He began studying Latin in the 10th grade in Brownton and continued with it through all four years of college.

The Brownton school at that time included only the first two years of high school, and Stakman finished his schooling there in 1900, when he was 15. For his junior year, he attended Humboldt High School in St. Paul, staying with his uncle, who was a minister. He must have dutifully attended church and listened to his uncle's sermons, but it left no mark on him.

One of his teachers at Humboldt High, Miss Fannin, was

> very good in botany. We had a lot of fun there collecting plants and identifying them. She was also quite good in history. There was a snappish Latin teacher, Miss Burlingame, who would have been called an old maid, and had some of the characteristics of one, but if anybody wanted to learn she'd help him in every possible way.

Because of his work in Brownton in the fall, he came to Humboldt a month late. Exams were given once a month, and he passed the next exam with no difficulty.

Before the next school term opened, in the fall of 1900, his uncle moved away from St. Paul, and for the first semester of his senior year Stakman attended Stevens Seminary in Glencoe. Stevens Seminary was a public school and later became Glencoe High School. Glencoe, the county seat, was about 12 miles east of Brownton. Stakman got the job of assistant janitor in the school and swept the floors and dusted the desks "with zeal and gratitude." At first he lived in the attic of the school, which must have been cold, dark, and drafty, with winter coming on, but later the janitor took Stakman into his own home.

The school gave a year's course in Latin in one semester. Stakman and a boy who became his lifelong friend, Michael O'Shaughnessy Higgins, tied for first place in speed of Latin translation without error—17 lines of Virgil per minute. This must have been fast translation by any standards, since it meant a line in just three and a fraction seconds, or several words per second. The school opened at 9:00 in the morning, but Stakman and some of his friends gathered there between 7:30 and 8:00 and quizzed one another on Latin, mathematics, and physics. The "boys hate school and teachers" theme, so common in popular American writing since the days of Tom Sawyer, certainly had no validity there, if indeed it ever had any validity anywhere among the better students. Stakman and his fellow students looked upon schooling as an opportunity, not as drudgery. Students like that must have been fun to teach, barring, of course, occasional confrontations regarding round and oval eggs.

The Glencoe school did not offer, in the second semester that year, the course in solid geometry that Stakman needed to satisfy the university entrance requirements, so he returned to St. Paul. He worked for his board and room with a family that lived on Crocus Hill, and he walked several miles each way to Cleveland High School. There he found the pace of translation of Virgil somewhat slower than the 17 lines per minute that he and his friend could handle at Stevens Seminary, since it was only 25 lines per day (presumably per class session).

Stakman graduated from Cleveland High in the spring of 1902 and returned to Brownton to work during the summer. He wanted to go to college and particularly to the University of Minnesota, being attracted in part by the reputation of the then president of the university, Cyrus Northrop, a "Downeaster" who had come to Minnesota with the avowed intent of developing culture in the crude Midwest and who had to some extent succeeded. Few or no university scholarships were available then. Stakman was offered a scholarship to the preparatory department of Carleton College, in Northfield, Minnesota, but preferred to go to the university if it was at all possible, a preference in which he was supported by his mother and sister. Tuition and fees at the university were $15 per semester. He worked during the summer and in the fall got a job on a threshing crew, which must have taxed him somewhat, since threshing in those days was hard and heavy work from dawn to dusk and sometimes beyond, and when he entered the university he weighed only 126 pounds. At 17 he must have had a tough and wiry body, as well as a tough and agile mind. He was eager to be off to the university to meet whatever challenges and opportunities it might offer. He could not then foresee, of course, that, with the exception of three years spent as high school teacher and administrator, his affair with the University of Minnesota would continue virtually unbroken for the next 75 years—a fruitful association that would endure almost to the end of his life.

CHAPTER TWO

College Days

After he paid his tuition and bought his books, Stakman had 65 cents left. He worked at various odd jobs to help pay his way and was able to make it, although at times it must have been nip and tuck, and he must have lived frugally, as have many other college students before and since. There were no formal "majors" then, but rather "group subjects," of which botany, German, and political science were three that he chose. His freshman courses included English, rhetoric, mathematics, botany, German, and Latin. He took both German and Latin all four years at the university

Concerning German, he says,

> German was another group subject. I had a lot of courses in German, took German all four years, plus a teacher's course in German, and some extra work in it, and I did that [not because of] the German language nearly so much as because I began to like German literature very much from the very beginning. I thought that some of their folksongs, some of their plays, some of their short stories were beautiful.

Of Latin,

> Then there was another man, by the name of [John E.] Granrud, in the sophomore year, who really made Horace and Tacitus, particularly, very significant, and the really important thing was that we studied Roman literature along with the language, all of the time, and so I would have to say to some of those old teachers: "I'm sorry that I didn't have sense enough to tell you this when you were still alive, but whatever good Latin has done me, I will always be grateful to you for having helped me to live vicariously in ancient Rome as long as I did."

In language and literature, Stakman was an internationalist long before he ever saw a foreign land. He also took enough French to be at least comfortable in it and considered taking Greek but decided against it.

Like all able-bodied male students at land-grant colleges and universities then, Stakman took two years of military, or ROTC. Once the rudiments of close-order drill were learned, the ROTC training then, as for many years later, involved mostly marching in squads back and forth across the dusty parade ground. Stakman says, "The first year I was a good soldier, but the second year I was a wounded soldier as much of the time as I could be wounded." He had one minor run-in with the military:

> The commandant was a major in the regular Army. He'd been in the

cavalry, and those fellows were supposed to be tough, and on general review of a Saturday when the band was out and all that sort of thing, and we were standing at parade rest, I heard this major say, "Take that grin off your face." I tried out my peripheral vision, looking to both sides to see who was grinning, but he pointed at me and said, "You! There!" I didn't know I was grinning. And so I went to his office and told him what had happened, and I told him that I thought I had a right to resent it, that I was not responsible for my looks and that I had not been intentionally grinning, that I'd been trying to be a good soldier. He talked a little rough for a while, and I told him that I would like to have him remember certain sorts of things. We talked about the thing for a while. Finally he said, "Well, I think if I had a face like yours, I'd disavow responsibility for it also. That's what we do in the army quite often." And so we parted good soldiers.

Maybe this is a fair and accurate, if not disinterested, account, but the possibility exists that Stakman actually was trying, with the trace of a grin subdued but still perceptible, to bait the commandant, succeeded in doing so, and was caught up for it. Stakman did not grin or grimace very much, ever, but when he did, it usually was intentional.

Botany and zoology as Stakman encountered them at the university "were a fairyland." In both of these subjects, the concept of organic evolution, which was developed by Charles Darwin in his *Origin of Species* in 1859 and which by 1900 was espoused by biologists throughout the world, was the central theme, and in both of these subjects Stakman had outstanding teachers. In biology,

[Charles P.] Sigerfoos was one of the clearest and most lucid lecturers that I've ever heard, and a very fine man. He had a phenomenal memory. They used to say that he'd never forgotten the name of a student that he'd had in his class. I saw Dr. Sigerfoos after I'd gone back to the university, after I'd been out about three years, and he met me and called me by name, and I don't think that it was because I was an outstanding student. I think it was because he remembered virtually everybody.

Stakman took the two-year or "long" course in botany. The lectures were given by Conway MacMillan, "a brilliant lecturer and a brilliant man, and as he lectured you could just sort of see the intellectual horizons expanding."

Dr. E. M. Freeman, who a few years later was to organize and head the Division of Plant Pathology and Agricultural Botany in the Institute of Agriculture on the St. Paul campus, was in charge of the laboratory work in botany.

At the end of the sophomore year, when I was working on my special problem [the anatomy of *Arisaema triphyllum*, jack-in-the-pulpit] he sat down and asked me whether I liked botany. I told him I did, and he almost scared me away from botany because he said, "You could be a good botanist if you wanted to be."

Another instructor in botany who impressed him was Carl D. Rosendahl:

Well, I must say also that along with MacMillan and Freeman in botany there was a man by the name of Rosendahl, from whom I took a year of taxonomy. I think probably he was the first professor with whom I felt completely at home. I'd felt pretty much at home with Dean Freeman, and I did with MacMillan, curiously enough. Maybe I was as tough as he was, I don't know, but I felt at home with him. But Dr. Rosendahl—we used to go out on field trips a good deal with the taxonomy class, and I mean, he was just like the rest of us. You'd get in class, however, and he was still the professor. He was a wonderful teacher in his way and a very, very fine man.

Dr. Rosendahl continued his outstanding teaching in botany for many years. He taught field taxonomy to the freshman foresters at the summer session at the university's Lake Itasca Forestry and Biological Station in northwestern Minnesota, and he is remembered with affection and respect by many of them.

Stakman's other field of specialization or "group course" (besides German and botany) was political science, which he also greatly enjoyed.

I think that political science was extremely interesting because of the subject matter and also because of the efforts of Dr. [William A.] Schaper, that old Dutchman, to really make political science function in the development of a consciousness of citizenship. In courses such as American government and comparative government and municipal administration and similar courses, he kept in mind all of the time not only the detailed organization of the government, the machinery of it, the method of operation, but also the theory and psychology back of it. I don't think anybody could have taken his course in comparative government without realizing that as compared with most other governments, that of the United States, with all of its faults, was so much better that it was worth preserving. I think we got the general idea of the amount of time, the amount of effort, and the amount of treasure and amount of blood that had been spent in getting certain basic rights and having them established, and I think this had a very great influence on all of the students.

Stakman also liked and was impressed by courses he took in the psychology and philosophy of education.

The philosophy of education was a wonderful course, not necessarily so much because of the lectures but because of the reading that you did in connection with it, although the lectures were among the smoothest that I ever heard. How profound they were, I'm not going to say right now, but the lecturer was Dr. George James, who was somehow connected with the famous James family.[1] ... Somehow or other from that time on I had the idea that education is the evolution of the inherent abilities and capacities of the individual, and I think that's terribly important. In all of the teaching that I've done, more and more I am convinced of the fact that the genes set the limits to what individuals can accomplish, and that what they actually do accomplish is the result of heredity and environment. And it seems to me that a great many mistakes are made in trying to push people into fields for which they do not have aptitudes. ... I probably got the idea too that education could do more than it can do. That may have been my fault that I got that idea. But I did come away sort of wondering what was going to happen when everybody had developed his intellect to the limit, and everybody had a gargantuan intellect. I was wondering who was going to dig the ditches and shovel the coal and feed the cattle. After I got out and had taught school for about six weeks, I quit worrying about that completely.

Stakman's extracurricular activities were pretty much standard fare for proper and well-behaved young men—lectures, concerts, plays, visits to museums, attending political rallies. After attending classes, studying, and working at odd jobs to earn money for his support, he cannot have had much time or energy left for hell-raising, even if he had been so inclined. If he and his fellow students had impure thoughts or lusted after forbidden fleshly pleasures, as young men will, no mention is made of it.

In the summers he worked at Jim Bohn's store in Brownton or on farms near home. He received the Bachelor of Arts degree in the spring of 1906. He was elected to Phi Beta Kappa, the undergraduate honor society of the arts college, but seldom wore the key, as some members did all their lives and some even in death. Nor did he ever wear a key indicating his membership in Sigma Xi, the honorary academic research society. He did not wear lapel buttons, rings, or insignia or baubles of any kind to indicate honors received or membership in select or at least restricted groups. He was always proud of his long affiliation with the University of Minnesota and was also proud of being a member in good standing of the human race, but he did not carry trinkets to show it.

CHAPTER THREE

Schoolmaster

After graduation, Stakman was offered an assistantship or fellowship in the Botany Department and also one in German,

> but I thought I'd better teach for at least a year and try to find out what my permanent interests were, and also to earn a little money and get even with the board. I had a little financial debt, and I had a very considerable debt to my sister and certain other people because they had kept on helping support the family while I was at least withholding my support. So I decided that I'd teach.

His first teaching job was at Red Wing, Minnesota, about 50 miles down the Mississippi River from St. Paul, in a beautiful setting of wooded hills and bluffs. The school had some problems with discipline, in part concerning athletics, which some of the students wanted to be an integral part of school life and which the authorities did not. Stakman convinced the authorities that athletics would be a good thing for the students and, in addition to his teaching (German, political science, history), wangled the job of director of athletics and coached the football and baseball teams. He must have fulfilled his duties and obligations to everyone's satisfaction, because he received the support of the school board, the superintendent of schools who had hired him, the faculty, and the students. After the school year was over, he returned to the university for the summer school, concentrating on history and mathematics.

> I've got to say that this summer was the first time that I had been able to study when and wherever I pleased, without having any outside work to do, and I enjoyed it thoroughly. I thought that maybe I could even become a bookworm, that the life of a scholar might be very pleasant, but there was something still pulling me, making me want to do something that was useful. . . .

The next year he taught at Mankato, in south central Minnesota. Mankato also had problems of discipline, and Stakman evidently had come there with a recommendation as one who could cope with such problems. He taught a variety of academic subjects (political science, history, algebra, botany) and coached football and track. Again he must have done an excellent job, because after the school year was over he was offered the superintendent position in schools in four towns in different parts of Minnesota. He was then 23 years old.

He chose to go to Argyle, a town of about 300 inhabitants in the Red River valley in northwestern Minnesota, about 50 miles south of the Canadian border;

there were still 10,000-acre "bonanza" wheat farms there. Besides being in charge of instruction, as superintendent, he also taught a heavy load of courses, including some that he organized himself to suit what he considered to be the needs of the community. He also organized and coached the track team, and at the regional meet in late spring his team made a higher total score than the three other competing teams combined. This went over big in Argyle, as did his more formal educational efforts, and at the end of the year the members of the school board, in a play to keep him there, offered to invest in the local banking business any money he could save. At that time, these investments were paying an annual return of 85 to 135%. Stakman gave up this alluring prospect of wealth and returned, at Freeman's renewed invitation, to the university to take graduate work in the newly organized (1908) Division of Vegetable Pathology and Botany (the name very shortly thereafter was changed to Division of Plant Pathology and Botany). In 1957, 48 years later, he wrote a somewhat nostalgic letter to an Argyle resident,[1] in answer to a request for his earlier impressions:

I am happy to give some recollections of the school year of 1908-09, when I was the proud 23-year-old superintendent of the Argyle school. In retrospect I am sure that I made plenty of mistakes but, like a baseball umpire, I did not admit all of them at the time. Nevertheless I learned a great deal from the townspeople and from the students and have always hoped that the students also learned something interesting and useful.

The first thing that impressed me was the strong desire of the school board and the townspeople to have a good school, even though there was not complete unanimity as to what constituted a good school. Certainly there was no unanimity as to what constituted a good superintendent; and I had moments of doubt myself.

There were times when I was forcibly reminded that the superintendent of schools was an important person, as he dealt directly or indirectly with most of the boys and girls of school age in the town. I recall a number of incidents in which I was reminded that no superintendent or teacher lived in a vacuum. Unfortunately my age became known to some of the townspeople, and I recall one old gentleman who expressed himself with more emphasis than elegance with respect to the inadequacy of a kid superintendent. In an attempt to use the laboratory method in teaching some features of physical geography, I had a class outside when the snow was melting, so that all of us could watch the formation of canyons, oxbow lakes, deltas, and various other physical features of the earth's surface in miniature. I recall the old gentleman standing on the sidewalk some distance away, shaking his head, and learned that he paid his disrespects to the school board for hiring a superintendent who played in the mud with the students during school hours.

The subject of communications—a euphemistic term for composition, rhetoric, and grammar—was acute then as now. In an attempt to see what could be done about improvement in oral and written exposition, we had a series of debates, some of which were open to the public and came on Bobby Burns Day. To the people of Scotch extraction this may mean something. It certainly meant something to me when a left-handed debater became very emphatic, gesticulating freely with his left hand with greater vehemence than coherence. This was the only time that it happened however, as the boys soon afterwards went into training for track. [I gather from this somewhat obscure sentence that the left-handed debater was so worked up he was threatening ECS with his fist, a not unusual impulse in opponents being worsted in arguments with him.]

In an attempt to improve spelling, we had periodic spelling lessons in connection with eighth-grade English. By mutual agreement we decided that each misspelled word would be written 10 times, with compound interest each successive time the word was misspelled. One day the students insisted that I had not assigned a spelling lesson, and I insisted that I had. Having had more experience in debate than they, I finally came close to persuading them that they had all been mistaken. Automatically some of them began to write words after school, and about 5:30 I found that a couple of the girls had already written the misspelled words about 2,500 times. I was horrified and apologized to them and sent them home, but this did not end the matter. In the evening a member of the school board came up and paid his disrespects to my lack of judgement in no uncertain terms. After he had cooled off somewhat, he informed me that a woman was at the bottom of it. This was not a new complaint! His wife just about had him convinced that such intellectual brutality required drastic measures, but he finally said, "Well, you know women." Thereafter, spelling assignments were written on the board.

On the theory that youngsters had a lot of superabundant physical energy and some of their intellectual energy might assert itself provided the physical energy was partly used up, we converted a basement room into a so-called gymnasium and equipped it with several sets of boxing gloves, a couple of wrestling mats, a handball court, and gave the boys a chance to use their energies against each other. Despite occasional sputterings of criticism about training prize fighters and grunt and groan wrestlers, the program bore fruit.

The track team in the spring was a wonderful demonstration of what a group of boys could do if they determined to do their best. Some of the boys had sprinted, but, so far as I know, none had ever put the shot, thrown the discus or hammer, or used a vaulting pole. Nevertheless, at the track meet of the Northern Red River Valley Athletic Association, comprising Hallock, Stephen, Warren, and Argyle, the Argyle boys made more points than all the other teams combined. In fact I think they made two and a half times as many points. At the start of the mile race, I was giving the Argyle participants some advice and ran along a little way to give them my parting word. The coach of one of the opposing teams threatened to protest the race, because he said that I was pacing the Argyle boys. As we turned to glare at each other, we discovered simultaneously that one of the Argyle boys was about a block away, and the opposing coach said, "I withdraw my charges. You couldn't pace those boys even if you wanted to; they can run and you can't." I tried to give the boys credit then, and I want to give it to them again. They trained hard, disciplined themselves, and accomplished what we set out to accomplish. Best of all, however, they kept sports in proper perspective, realizing that, after all the primary business of the school was intellectual development and that sports had merely supplemental value as recreation and self-discipline.

There are more evidences of self-expression. Looking out of the windows during morning exercises it was interesting to count the number of people living near the schoolhouse who either opened or closed their windows, depending on whether or not they appreciated the lustiness with which the youngsters sang Dixie, Rule Brittania, and certain other songs that lent themselves to volume as well as melody.

Then, as now, there were curricular problems, especially as the school at that time was relatively small. Vocational education was in its infancy, and about all that could be offered was a course in bookkeeping, and some elementary agriculture in botany. The people of the town were not in complete agreement with respect to the principal functions of a school, but all of them agreed that one of the functions should be to help give the students a basis for intelligent citizenship. Attempt was made to do this, but the efforts sometimes boomeranged, as when some of the students argued that constitutional rights and certain freedoms should be given to the students as well as adults. Even the arguments, however, had some value.

The greatest compliment that I could pay the students in the school, the members of the school board, and the townspeople generally was that they wanted an efficient and orderly school, and most of them were willing most of the time to concede that there were multiple objectives for multiple students. For the most part the students disciplined themselves, although occasionally there was a little eruption that had to be quieted with a little show of authority. It is a fair conclusion, however, that 98 percent of the students and the townspeople were reasonable 98 percent of the time, if they could catch the teachers in an equally reasonable mood at least 85 percent of the time.

All in all, I want to pay tribute to Argyle as a community of fine people who wanted a good school, who did everything that they could to help make it a good school, including the students themselves. Now that Argyle is to be congratulated on furnishing better physical facilities, I am sure that they will never forget that the primary objective of a school is to give the students, each according to his aptitude and industry, the best possible opportunity to make the most of his education, so that he can lead the most useful and rewarding life that is possible for him.

Nowhere does Stakman so much as mention any of his fellow teachers in the three schools at which he taught. Staff meetings must have been held occasionally to discuss problems and procedures, and at Argyle, where he was superintendent, he must have presided over these meetings; teaching, especially in a small-town school, is necessarily a joint effort. There must have been support, and opposition, among the other teachers on the staff, some differences of opinion as to goals and methods of reaching these goals, some pushing and pulling, some sharing of burdens; Stakman could not have done all this absolutely alone. Yet nowhere does he give the slightest acknowledgment or recognition of any joint effort. Much later on in the oral history, he mentions very briefly a few of the staff members who shared with him to some extent some of the responsibilities of teaching, research, extension, and even administration during his 40 years as head of Plant Pathology at the University of Minnesota, but it is no more than a token mention. He was on center stage; they were in the wings. This was not true of his associates in the Rockefeller Foundation's Mexican Agricultural Program, nearly all of whom were on center stage with him. This may not be important; it's just the way it was.

CHAPTER FOUR

Graduate Work

After the year at Argyle,

> I decided to go to the University of Minnesota for graduate work, partly as a result of desire and partly as a result of opportunity.... I had and declined offers of a scholarship in botany and an instructorship in German, and had been encouraged to do graduate work in political science at the university.

The instructorship in German carried the unusual option of his being able to do graduate work in any department he chose. He accepted Freeman's offer of an instructorship in plant pathology for two reasons. First, "I wanted to study with Dr. Freeman if I could." Second, "I wanted to be as scholarly as I could but I wanted my scholarship to be productive. In other words, I wanted to combine knowledge and interest with utility as much as possible." In still other words, he wanted to do well for Stakman, but he also wanted to do good for his fellow humans, and for this, at this time, he probably could not have selected a better field than plant pathology in which to develop and use his talents.

The idea of being able to acquire useful knowledge in plant pathology was partly a product of his own experience. In 1904 there had been a severe epidemic of black stem rust in Minnesota and neighboring states, and the farmers around Brownton had lost from 50 to 100% of their wheat. Stakman was in college then but worked in Brownton and on farms near there in the summer. In one of his accounts, he "stooked" 200 acres of wheat that fall (that is, gathered up the bundles of wheat bound by the binders and stood them up in stocks—in the Midwest it was called "shocking," but he was talking to an audience in England when he referred to it as "stooking"), which would have been quite a task for a young man, however vigorous. In another account, he clerked in the village store and was quizzed by the farmers about the cause of the rust epidemic, since he had been to college and supposedly was educated. Maybe he did both—it was a common practice for a young man to quit whatever job he had at harvesttime and join a harvesting and threshing crew, where the work was harder but the pay was better. In the meantime,

> I went back to Brownton after getting through with Argyle in early June, and I repaired the roof just to show that a person who had gone to college and was teaching school could do a better job of it than the tinsmith who'd tried to stop a leak in the roof, repaired a wooden sidewalk, and grubbed out a bur oak, which has a taproot and is hard to grub out [the size of the bur oak was not specified, but a large bur oak has a really massive root system that

even Stakman could hardly excavate], then I was going to stop and think a little before going down to the university. But I got a letter from Dr. Freeman—the weeds were running away with the experimental plots, could I arrange to come down a little early? So I went down a week or ten days before the appointment started and went to pulling weeds, controlling weeds, and finally after doing that for a while, my appointment started on July 1, and I became instructor in vegetable pathology, now called plant pathology, and assistant in the experiment station. But I was instructor in plant pathology without ever having had a course in plant pathology or in any other agricultural subject, for that matter, but I had had a lot of experience and knew quite a little about agriculture from the practical standpoint.

Stakman was not given to doubt about his competence at any job or any activity he undertook.

At any rate, I had never had these courses but I soon had something that was of infinitely more value, in my opinion, than any courses could have been. That was the opportunity to be associated with Dr. Freeman and the USDA (United States Department of Agriculture) collaborators and a few other people in the actual work in the experimental field plots, and later I had the inestimable and incalculable privilege of helping Dr. Freeman teach his courses. I learned plant pathology by helping to teach it, and I also had the privilege of very soon doing some independent research.

Since Dr. E. M. Freeman provided the stimulus and the opportunity for Stakman's entrance into plant pathology, he deserves a brief description. He was Stakman's first laboratory instructor in botany, then his advisor and preceptor, and later, his long-time associate in research and administration. Freeman received his M.S. and Ph.D. degrees at Minnesota, majoring in botany but with strong leanings toward plant pathology, in which there was then no formal instruction anywhere in the United States. He did postgraduate work with Marshall Ward at Cambridge University in England. Marshall Ward was a major prophet in plant pathology at the time, and his every pronouncement was regarded pretty much as revealed truth. Like most other mortals, he sometimes was wrong, and one of these times concerned the nature of the stem rust fungus, with which Stakman was to have a lifelong affair.

Freeman was a very able and productive scholar and researcher. In 1905, he published a book, *Minnesota Plant Diseases*,[1] of 432 pages, illustrated in part with excellent photographs taken by himself. It described the major diseases of economic plants in the midwestern United States and was the first textbook on plant diseases in the United States. Although it is outmoded now, at the time it was an excellent compendium of basic and practical information on the nature, cause, and control of plant diseases, and some of the material in it still is useful. Freeman and others recognized the need for much more work in plant pathology, a need dramatized by the disastrous epidemic of stem rust in 1904.

Freeman went to Washington, D.C., to work with USDA plant disease experts and to instigate collaborative studies with the USDA. After his return, he helped organize the Division of Vegetable Pathology and Botany, on the St. Paul campus, which was formally established August 1, 1907, with Freeman as its head, the first such division or department in a university in the United States. In

1913, Freeman relinquished direction of the Section of Plant Pathology to Stakman but until 1937 continued as administrative head of the Division of Plant Pathology and Agricultural Botany. He continued to give occasional lectures in mycology until the early 1930s, but these were devoted to rather abstruse aspects of the evolutionary derivation of some groups of fungi from some groups of algae, a system of the course of evolution developed in the 1880s, now long since out of favor and even in the 1930s of doubtful validity. Moreover, since few or none of the students taking mycology at that time had had even the slightest experience with the algae that loomed so large in this esoteric discussion, Freeman's sometimes two-hour-long lectures left them somewhat dazed.

On field trips, however, Freeman was delightful. On Saturday morning, four or five graduate students especially interested in trees, shrubs, fleshy fungi, and other outdoor flora would lay in a supply of choice sirloin steaks (at about 15 cents per steak) especially cut for broiling over an open fire, pack other edibles, and drive off with Dean Freeman to a forested area near Afton, on the St. Croix River. They would spend the day botanizing and mycologizing, with a long break in early afternoon to broil the steaks and sit around the fire discussing. Freeman had a tremendous knowledge of botany and a very keen mind, and he did not throw his academic weight around but was just one of us. For getting acquainted with living things and their complex interrelationships, field trips such as those, with a man like Freeman, were incomparably superior to any formal courses. They have long since gone out of style.

When Freeman offered Stakman an assistantship in plant pathology, with the opportunity to work for an advanced degree and at the same time serve an apprenticeship in this new and exciting field, Stakman accepted with alacrity. At that time, in 1909, only two men on the St. Paul campus had Ph.D. degrees; one of them was Freeman. Stakman was the first graduate student on the campus, and of course the first one in plant pathology. Graduate work was not so highly regimented then as it became later when insistence upon credits and grade point averages replaced emphasis on learning and research. Stakman majored in plant pathology and helped teach the advanced course in it taught by Dr. Freeman. He minored in ecology and entomology. The graduate school was formally established during 1909–1910, with Professor Henry T. Eddy, from engineering, as dean. When Stakman finished the "required" course work and thesis for the degree, he was asked by Dean Eddy, "Did you study particularly for technical purposes, or technological purposes, or for the purpose of an education?"

> "Well," I said, "since I'm not completely sure that I can be a professional, I guess you'd have to say I studied particularly for educational purposes." He said, "Then you can have a Master of Arts degree if you want it." I said, "Give me a Master of Arts degree. Okay, I'll take that degree."

Graduate education was considerably simpler then. Throughout his tenure as head of the plant pathology department, Stakman did his best to keep it simple for students majoring and minoring in his department. He did not require a certain number or sequence of courses and, in the courses he taught, he never gave any grade other than S (satisfactory), which for years the clerks in the graduate school office translated into B.

By the time he received the M.A. degree, Stakman, with his mastery of several languages and his wide reading in diverse fields, was already well on the way to becoming an educated man, a well-rounded scholar.

Stakman's thesis work for the M.A. degree dealt with germination of spores of the fungi that cause smuts of cultivated cereal plants in Minnesota. The smut fungi had been investigated fairly thoroughly by European mycologists and plant pathologists, and the standard germination patterns had been described and illustrated. Stakman found, however, that some of the smut fungi he worked with did not behave as they were supposed to; their germination patterns differed from that shown in research papers and books. From their own work, Freeman and Stakman soon began to realize that the fungi they worked with were variable. That is old stuff now, but it is old stuff partly because variation and variability in fungi continued to be the general background, and often the foreground, theme of Stakman's work for the next 40 years.

At the time he began his research with smuts and rusts, in 1909, all species of fungi were supposed to be fixed in an unchanging pattern. "Species are fixed" was the dogma of the day so far as fungi were concerned. Mycologists and fungus taxonomists distinguished one species from another on the basis of the slightest possible differences, with no knowledge of or concern about environmental or genetic variation in the character or characters chosen. This was a hangover from pre-Darwin times; one of Darwin's great contributions was the general principle that all species are varying in all characteristics at all times and everywhere. Work in population genetics since about 1930 has gone a long way to explain the dynamics of speciation, but Stakman in 1910 soon recognized that, as far as the fungi he was studying were concerned, species were man-made. One of his important contributions to biology was the establishment of the principles of variation and variability in fungi and in microorganisms in general.

During Stakman's long tenure as head of the Department of Plant Pathology at the University of Minnesota, most of the research of most of the candidates for advanced degrees dealt with the problem of variation in one form or another. They examined its mechanisms and meanings, and especially the development, by hybridization and mutation, of new types. Darwin's success in establishing the principle of evolution was due in part to the tremendous mass of supporting data he accumulated over years of study. Similarly, Stakman's success in establishing the principle of variability in fungi was due in part to the tremendous amount of data accumulated with many different kinds of fungi over a period of decades. This principle or idea of variability, as it applied to parasitism by rust fungi, was difficult to get across to some plant breeders, even up into and beyond the 1930s; they thought that once they had produced a variety of wheat resistant to the then-known races of stem rust, they had a variety permanently resistant to all races of stem rust. One of Stakman's famous phrases, coined in the 1930s, was "Plant diseases are shifty enemies." He was already seeing this in his work with the germination of smut spores in 1909–1910. The implications were only dimly perceptible, but the germ of the idea was there, and over the years he developed it into an all-pervasive principle of universal application.

In his work for the Ph.D. degree, Stakman considered going to some place other than the University of Minnesota, to avoid the possible taint of "inbreeding." (During his years as head of the department, he hired as staff members almost only those who had received most or all of their postgraduate training in his department, although most of them had had some postgraduate experience elsewhere.) One limitation on graduate study in plant pathology elsewhere, at the time, was that there was only one other department of plant pathology in the United States. It was at Cornell University and had been established shortly after the one at Minnesota, with Dr. H. H. Whetzel as head. Stakman visited there and decided that Cornell offered no advantage over Minnesota. However, at the time, the USDA in Washington, D.C., had a tuition-free graduate school, with a large group of experts in plant pathology and related agricultural sciences. Stakman spent the winter of 1911–1912 there, working in the laboratories of Erwin F. Smith, pioneer plant bacteriologist, and K. F. Kellerman, soil microbiologist. He was offered jobs by both men, the one with Kellerman paying twice as much as he was getting at Minnesota, but he refused the offers. He also circulated freely among the "big shots" in plant pathology at the USDA in Washington and established working relationships with some of them. He considered it a very worthwhile experience, and he doubtless soaked up a lot of information. He returned to Minnesota in the spring of 1912 and continued his graduate work for the Ph.D., completing it in the spring of 1913.

One of the requirements for the degree was knowledge of two foreign languages, German and French. The examination in German was more or less a formality, as it should have been, since a few years before he had been offered an instructorship in that department. However, he dutifully displayed his familiarity with German by spouting passages from Goethe's "Faust" (which he could and occasionally would do many years later).

The examination in French was somewhat different, although also oral. Charles W. Benton, the professor of French who was examining him,

> gave me an examination first in plant pathological French. He said, "Well, you ought to know that. That's in your field and that doesn't indicate that you know the language very well." So he tried something else, and that was on terrestrial magnetism, and fortunately there were certain terms in that book that I knew because of a knowledge of geology, that he didn't know. So he decided that vocabulary wasn't the most important thing, that, after all, a knowledge of grammar was, so he picked out something literary and he asked me all the grammar he could. "And what mood is that in?" "The optative subjunctive." "What do you know about the optative subjunctive?" And, "What kinds of subjunctives are there?" So he gave me a good examination in French grammar, and he said, "Well, just because a person knows the grammar, he doesn't necessarily know the language." And so he asked me what the cognates were, the equivalents, the derivation of words. Well, fortunately I'd had eight years of Latin, and many of the French words of course came from Latin. Then that wasn't quite enough, the next one was, "What English words came from them?"

This went on for some time. Maybe Professor Benton was only brightening up an otherwise dull day by putting an obviously capable candidate through the hoops. Or maybe he was taking advantage of the opportunity to be a little sadistic. Whatever the case, this made enough of an impression on Stakman that, when he himself came to advise graduate students not too long afterwards, he led a reform movement to change the requirements to a capable reading knowledge of German and French. He also did his best to make sure that candidates in plant pathology who came up for advanced degrees would pass the exams—"we had a record for years and years of never having a failure over there" (that is, in the French and German exams). Probably few of those who for "years and years" passed the exams in written French and German continued their study of these languages to the point where the languages were part of their scientific weaponry, but some of them did, and for these, Stakman's insistence upon acquiring a good basic knowledge of the languages was very worthwhile. The foreign language requirements for advanced degrees in scientific fields have long since been given up, for better or worse. Stakman felt that some competence in foreign languages was an integral part of scholarship, and he fought valiantly to retain the foreign language requirements for advanced degrees, but he eventually lost.

Stakman's oral exam for the Ph.D. was not at all fearsome. He had been in the hospital with an attack of pleurisy and was recovering. The exam was given in the hospital yard

> on a nice sunshiny day under the trees, by Dr. Freeman, who was chairman of the committee, and Professor [Frederic E.] Clements, professor of botany, and Professor [Alois F.] Kovarik, of physics, and I would say that that was a model Ph.D. examination. They asked me to tell in my own words about the accomplishments of my thesis, particularly, and the explanations for the results, insofar as they were available, and the significance, and they interrupted occasionally to ask a question, but they wanted to know what I had accomplished. Then they asked me some questions, but they also asked me this—"Now, what new researches would you project?" And when I talked about epidemiology, they said, "All right, if you had $100,000, what would you do in order to try to solve this problem of the origin and distribution of rust and its relation to resistance?"
>
> Well, apparently I spent the $100,000 to their satisfaction, because they passed me. But I'm not saying that this was a good examination from my standpoint [he means that he didn't think he did exceptionally well]. They didn't try to embarrass me, they tried to find out what I'd accomplished and what I thought, was I a professional or wasn't I? I'm not saying that my performance was distinguished. If it hadn't been satisfactory I could have attributed it to the aftermath of the pleurisy in any case. But it seemed to me that, again, this examination was an indication that these people were educators and they wanted to see what degree of progress a person had made in higher education, and as scientists they wanted to find out what the prognostication was for producing new knowledge and not just simply rehashing old knowledge, and I'm very grateful to them; and in my more mature years, I've tried to keep that in mind also with students—that they are under the gun in an oral examination. Many of them never have had an oral examination before they come up for the "prelim" for the Ph.D., for example. Then they get a number of people there who act as if it were the Spanish Inquisition, without ever having prepared them for it, and this I think is cruel.

Stakman very soon was on the examiners' side of the railing, and, as a partial strategem to cope with the Spanish Inquisition syndrome in Ph.D. examinations, he and his then associates instituted what soon became standard practice, a preliminary oral examination. This supposedly tested a candidate's academic and intellectual capabilities in the major and minor fields of study and in one or more fields outside his specialty. Theoretically, it also helped prepare the candidate for the ordeal of the final oral examination. In practice, the latter benefit probably seldom was realized, any more than one stretching of a heretic on the rack in medieval times prepared him for a second stretching.

However, Stakman went much further than this with his own graduate students; they were subjected to multiple examinations by their fellow graduate students and by staff members, in weekly seminars and to some extent in informal discussions. A graduate student who was a candidate for the Ph.D. degree did not come up for a preliminary exam until Stakman was convinced that the student had a reasonably good chance of passing. All his graduate students first earned, or at least received, an M.S. degree before they were permitted to go on toward a Ph.D. If, in Stakman's opinion, students were not of Ph.D. caliber, they were eased out with an M.S. degree, with their self-respect intact, and were not destroyed by flunking a Ph.D. exam. It was a humane and civilized approach. This could not, of course, stop all petty and deliberate torturing of candidates such as some examiners delight in, but it enabled the candidate to cope more effectively with it, if in no other way than that, after having lost all arguments with Stakman for the better part of three or four years and having been driven to deep study and such hard thinking as graduate students are capable of as a result of these arguments and discussions, the student was less likely to be scared and scarred by the cross-examination of a less competent examiner.

Stakman ended his discussion of examinations in the oral history with these comments:

> ... when anybody ever asked me, "Has a certain individual had his preliminary examination?" I'd say, "Yes, he's had a lot of them. He's had a preliminary examination once or twice a week at least." Because I mean all the way through you give a person an examination without his knowing that he's taking an examination. The best examinations that I've ever given anybody was just to talk with them completely informally in a friendly way about what he was doing and what he knew and exchange ideas with him. That's the best kind of examination.

One reason why most of Stakman's students did not get worked over so roughly as the students of some other advisors in other departments was that in the preliminary and final oral examinations Stakman dominated the questioning, sometimes, in fact, to the point where the other members of the examining committee, and sometimes even the candidate being examined, were almost completely shut out. Stakman simply took over for two or three hours; he also furnished a supply of choice cigars (professors still smoked cigars, then), several brands of cigarettes, coffee, and assorted goodies and, had the rules permitted, probably would have served wine and maybe even martinis, to mellow the mood. He made the occasion, if not exactly festive, at least a gentlemanly gathering of scholars, as civilized as such an ordeal could be. After two to three hours of domination by Stakman, only the meanest and most insistent of examiners would spend more than a token amount of time on the candidate, and even then Stakman would make long parenthetic interjections and discourses. Once in a while he let one of his more uppity students suffer on the rack for a time, to sort of tone him down for his own good, but this was done with good-humored benevolence, not malice. There was no meanness or malice in Stakman's nature; as will be evident later, he might sometimes rough up an opponent in the handball court and leave him somewhat dazed and shaken, but this was purely a sporting maneuver, without any malice whatever, and he bore no grudge against the opponent he had roughed up.

PART II
Plant Pathologist

Stakman, sometime in the 1920s.

CHAPTER FIVE

The Epidemiology of Stem Rust

Assistant in Plant Pathology
to Section Head, 1910–1914

Stakman received his M.A. degree in June 1910 and either was then appointed or already had been appointed Assistant in Plant Pathology, with responsibilities in undergraduate and graduate teaching, research, and extension. The content and format of his early publications give interesting information about Stakman and his colleagues at that time. His first publication, of which he was coauthor, was a bulletin, *The Smuts of Grain Crops.*[1] The bulletin was illustrated with some good pen-and-ink drawings of the different smuts described in the text. Whether these were drawn by Stakman or someone else is not known, since no credit is given or taken. However, in the oral history, Stakman bragged that as a student in botany, in one of Freeman's classes, he was very good at drawing plants, and it certainly was his practice to give credit to others when it was due, so maybe the drawings are his. A sample is shown in Figure 1. The bulletin must have been written in the fall of 1910.

The next publication of which Stakman was an author was *Orchard and Garden Spraying,*[2] a bulletin put out by the Divisions of Entomology and Vegetable Pathology. The cover has a photo of a horse-drawn sprayer, with the spray liquid in a barrel and the spray pump operated by hand, which must have required a strong, durable, and devoted pumper. Inside the cover, the experiment station staff members are listed (Figure 2). Freeman was the only one with a Ph.D. degree, but two men in Veterinary Science had the D.V.M. degree. These gave a slight aura of scholarship to what then and for many years afterward was officially known as the University Farm, somewhat sneeringly referred to by the erudite and nonerudite academicians on the Minneapolis campus as the Cow College. Stakman, who was to become one of the world's leading agricultural scientists, always was proud of his association with agriculture and agriculturists.

His next publication, *Potato Diseases,*[3] came out in November 1912, with Stakman as the first author. His coauthor, A. G. Tolaas, later was in charge of seed potato certification for many years until his retirement, when that activity was taken over by the State Department of Agriculture. Even in the 1910–1915 era, the growing of potatoes for seed, to be shipped to growers in states farther south for planting, was a thriving business in Minnesota, and it required some scientific aid and expertise.

The station staff are listed in the front of a 1913 bulletin, and there were then two men with the Ph.D. degree on the campus, Freeman and L. D. H. Weld, Agricultural Economist; even then, the "dismal science" was poking its head into the tent of agriculture. The Division of Plant Pathology and Botany

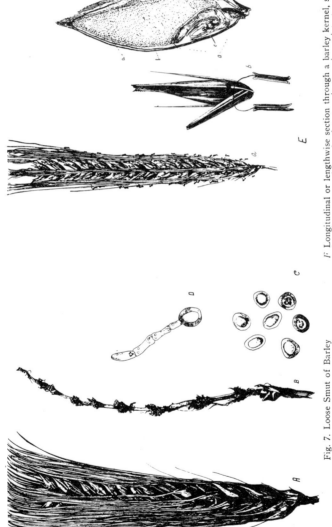

Fig. 7. Loose Smut of Barley

A Normal head of barley.
B Smutted head of barley.
C Smut spores; greatly enlarged.
D Spore germinating; greatly enlarged.
E (a) Head of barley, showing open glumes at the stage when the smut spores are probably blown in. (b) A single flower.
F Longitudinal or lengthwise section through a barley kernel, showing the young embryo plantlet at a, its growing-point at c; food for the young plantlet at d, seed coat shown at b. The smut threads live in the young plantlet a during the winter, and get into the growing point c while it is still inside the seed. (For detail of threads in growing-point, see Figure 8. Compare with Figure 6, noting difference in the stage at which infection takes place.

Figure 1. Sketch, presumably by Stakman, in his first publication.

30

(no longer Vegetable Pathology) contained two sections. One was Plant Pathology, consisting of Stakman, Assistant Plant Pathologist, In Charge; E. Louise Jensen, Mycologist (in 1917 she became Mrs. Stakman); and A. G. Tolaas, Assistant in Bacteriology. The other was Agricultural Botany, consisting of W. S. Oswald, Assistant Agricultural Botanist, In Charge, and Robert C. Dahlberg, Seed Analyst.

The next paper that carries Stakman's title is "Relation between *Puccinia graminis* and Plants Highly Resistant to Its Attack,"[4] which was published in June 1915. Stakman is identified as Head of the Section of Plant Pathology and Botany, Division of Botany and Plant Pathology, Department of Agriculture, University of Minnesota.

Stakman had been made Head of the Section on July 1, 1913, shortly after receiving his Ph.D. degree. So since 1910, when he received his M.A. degree ("Okay, I'll take the M.A."), he had progressed from Assistant in Plant Pathology, the lowest academic ranking, to Assistant Plant Pathologist, In Charge, to Head of the Section, the rank he was to retain until 1940, when he became Head of the Division of Plant Pathology and Botany. His rise through the academic ranks may not have been meteoric, but it certainly was rapid. As Head of the Section, he was in charge of most of the undergraduate and graduate instruction (and personally did a good share of it), most of the research (including field plot work and the planning of ongoing projects in cooperation with other departments and with the USDA), extension work (always high on his list of priorities), the advising of graduate students, and the routine administrative work with the graduate school and the experiment station. He was also involved in the development of undergraduate instruction throughout the College of Agriculture, about which he had some pretty strong opinions that he liked to expound at considerable length to the captive and sometimes restive audience at the monthly faculty meetings. For many years he was the principal speaker at these meetings, and sometimes the only one, since by the time he was through with his peroration, the afternoon classes were well under way.

Thus by July 1, 1913, six weeks past his 28th birthday, Stakman was well established in what was to be his lifetime profession and was already in a position of leadership, a position natural to him and one that he maintained for the rest of his life. The man, the time, and the place had come together. From then on he was a professional plant pathologist, with most of his efforts devoted to teaching, research, administration, promotion of support for plant pathology, and expounding the gospel of science to the public. Most of his research was with the epidemiology of stem rust.

Epidemiology of Stem Rust

Stakman's work with stem rust of cereals, which was to occupy so much of his time and effort for the next 40 years or more and carry him to international renown, began with his Ph.D. thesis, entitled "A Study in Cereal Rusts—Physiologic Races." This was completed in June 1913 and was issued immediately in the form of a bulletin but with no official university imprint of any kind. It consisted of two parts: I, Biologic Forms and II, Rust-Resistant Varieties of Wheat, and comprised 54 pages, including a bibliography of about 80 papers by 60 authors in five languages. (There is no reason to suppose that he

UNIVERSITY OF MINNESOTA

THE BOARD OF REGENTS.

CYRUS NORTHROP, LL. D The President of the University.	Ex-Officio
The HON. JOHN LIND, Minneapolis The President of the Board.	1914
The HON. ADOLPH O. EBERHART, Mankato The Governor of the State.	Ex-Officio
The HON. C. G. SCHULZ, St. Paul The State Superintendent of Public Instruction.	Ex-Officio
The HON. W. J. MAYO, Rochester	1913
The HON. MILTON M. WILLIAMS, Little Falls	1913
The HON. HENRY B. HOVLAND, Duluth	1914
The HON. A. E. RICE, Willmar	1915
The HON. CHARLES L. SOMMERS, St. Paul	1915
The HON. B. F. NELSON, Minneapolis	1916
The HON. PIERCE BUTLER, St. Paul	1916
The HON. CHARLES A. SMITH, Minneapolis	1916

THE AGRICULTURAL COMMITTEE.

The HON. A. E. RICE, Chairman
The HON. MILTON M. WILLIAMS
The HON. B. F. NELSON
The HON. H. B. HOVLAND
The HON. PIERCE BUTLER
President CYRUS NORTHROP

STATION OFFICERS.

A. F. WOODS, A. M.	Director
T. L. HAECKER	Dairy and Animal Husbandry
M. H. REYNOLDS, B. S., M. D., D. V. M.	Veterinary Science
ANDREW BOSS	Agriculture
FREDERICK L. WASHBURN, M. A	Entomology
F. M. FREEMAN, PH. D.	Botany, Plant Pathology
JOHN T. STEWART, C. E.	Agricultural Engineering
RALPH HOAGLAND, B. A.	Agricultural Chemistry, Soils
J. M. DREW	Poultry
C. C. LIPP, D. V. M.	Assistant in Veterinary Science
C. P. Bull, B. Agr.	Assistant in Agriculture
D. A. GAUMNITZ, M. S. Agr.	Assistant in Husbandry
A. D. WILSON, B. S. in Agr.	Assistant in Agriculture
LEROY CADY, B. S. in Agr.	Assistant in Horticulture
E. G. CHEYNEY, A. B.	Assistant in Forestry
A. G. RUGGLES, M. A.	Assistant in Entomology
THOMAS P. COOPER, B. S. in Agr.	Assistant in Agriculture
W. H. FRAZIER, B. S.	Assistant in Soils
A. R. KOHLER, B. S. A.	Assistant in Horticulture
W. F. HANDSCHIN	Assistant in Animal Husbandry
J. P. WENTLING, A. M.	Assistant in Forestry
GEORGE P. GROUT, M. S.	Assistant in Dairy Husbandry
W. L. OSWALD	Assistant in Agricultural Botany
A. C. ARNY, B. S. in Agr.	Assistant in Agriculture
R. M. WEST, B. A.	Assistant in Chemistry
D. P. TIERNEY, M. F.	Assistant in Forestry
B. A. BEACH, D. V. M.	Assistant in Veterinary Science
JOHN SPENCER, D. V. M.	Assistant in Veterinary Science
E. C. STAKMAN, M. A.	Assistant in Plant Pathology
H. B. SCAMMELL, B. S.	Assistant in Entomology
A. J. McGUIRE, B. Agr.	Superintendent, Grand Rapids Sub-Station
E. C. HIGBIE, M. A.	Superintendent, Morris Sub-Station
C. G. SELVIG, M. A.	Superintendent, Crookston, Sub-Station

Figure 2. The station staff, from a 1911 bulletin.

did not read the originals of all these—he did not have to bluff where scholarship was concerned, not that he was above bluffing if he thought it would be effective.) There were nine plates, seven of them of rusted seedling leaves of wheat, none of which show very much. The thesis was republished in February 1914 as Agricultural Experiment Station Technical Bulletin 138, with no word changes from the first version, indicating that Stakman and others must have been satisfied with the way it was presented in the first printing.

Part I, Biologic Forms, dealt with the then-recognized "forms" of the stem rust fungus, *Puccinia graminis*. These were *P. graminis tritici*, wheat stem rust; *P. hordei*, barley stem rust; *P. secalis*, rye stem rust; and *P. avenae*, oat stem rust. (Later, the barley stem rust form, *P. hordei*, was dropped, and another form, timothy stem rust, *P. phleipratensis*, was added.) Stakman was concerned mainly with the ability of each of these forms to infect hosts other than the one for which it was named, with the constancy of the forms' parasitic capability, with their ability to change or adapt by means of "bridging" hosts, and with the possible ability of anesthetics and fertilizers to alter the resistance or susceptibility of the hosts and the parasitic capabilities of the rust fungus.

To dispose of the latter, and minor, things first, he found no evidence that anesthetics had any effect on altering the parasitic capabilities of the rust fungus. This was not his idea but one that had been put forward by older but, as it turned out, not wiser, pathologists. Even if it had been valid, it would have been strictly a laboratory phenomenon of no conceivable practical significance—thesis biology. Stakman never had much time for strictly laboratory phenomena. In any case, he found no evidence that anesthetics affected the parasitic capabilities of the fungus, and nothing more was heard of it. He found that fertilizers might affect the severity of infection by the rust fungus but did not alter the parasite's inherent capability to infect a host.

The question of bridging hosts will be referred to in more detail shortly, but in his Ph.D. thesis work he found no evidence to support this concept. Even when the rust fungus went through the sexual stages on barberry, its parasitic capabilities were unchanged. *P. graminis tritici* remained *P. graminis tritici* and did not change into *P. graminis secalis* or *P. graminis avenae*. As one indication of the thoroughness of his work, he grew one collection of rust from wheat on einkorn for 19 months, with transfers every three weeks, then compared its virulence on einkorn and on wheat with that of the same collection of rust maintained only on wheat. They differed somewhat, but that might have been caused by selection, through repeated transfer, from a mixture of biotypes present in the original collection on wheat. This possibility was not explicitly expressed in the thesis, but it must have been present in his mind because it was an idea he was soon to explore very thoroughly and perceptively. The role of barberry in the production of new races of *P. graminis tritici* was to come out later. At that time, stem rust on wheat, *P. graminis tritici*, was thought to be a uniform and constant entity (except when it was changed into *P. graminis secalis* or *P. graminis avenae* or whatever, by bridging hosts), but Stakman was soon to change that, too.

In Part II, Rust-Resistant Varieties of Wheat (even then, Stakman's eye was very much on the ultimate goal, basic knowledge that would permit the rational development of rust-resistant varieties of wheat), he reported that varieties differ in severity of infection and in length of incubation period (the time from inoculation of the leaves to production of urediospores). The incubation period was seven days on Minnesota 163, 12 days on Arnautka 288, and 14 days on Khapli. The importance of "slow rusting" in the buildup of epidemics was not recognized, or at least not exploited, until much later.

Some of the work done by Stakman for his Ph.D. thesis was essentially a repetition of work done by others before him, some of it by rather famous men, at least as fame was then reckoned among plant pathologists and botanists. But insofar as bridging hosts altering the infection capabilities of the stem rust fungus was concerned, his results and conclusions were almost diametrically opposed to theirs. His work was more perceptively planned, more extensive, and more carefully done than the work of his predecessors; the results were more clear-cut; and his conclusions were sound. That they differed from some of those of the established authorities did not bother him in the least; more likely, he relished it. He was justifiably confident of his work and his results, and he enjoyed setting people straight.

His thesis also included some work on hypersensitive reaction in plants that were immune or almost immune to infection by *P. graminis*. In this situation, germ tubes of the germinating urediospores penetrated normally into the host tissues but went no farther because the surrounding host cells died and the fungus could not continue to develop. He followed this up in his next paper,[5] which has a summary that is a model of brevity:

(1) When plants practically immune to *Puccinia graminis* are inoculated, the fungus gains entrance in a perfectly normal manner.

(2) After entrance the fungus rapidly kills a limited number of the plant cells.

(3) The fungus, after having killed the host cells in its immediate vicinity, seems unable to develop further.

(4) The relations between plant and parasite in partially resistant and almost totally immune plants are different in degree only.

(5) Hypersensitiveness of the host seems to be a common phenomenon not only among plants somewhat resistant to *P. graminis* but also among those almost totally immune to it.

Not many wasted words there. Whether he received any editorial help on this is not known, but it seems unlikely, even given the almost uncontrollable urge of editors to improve manuscripts. If Stakman was not then the finished writer he eventually became, he knew what he wanted to say and he could and did say it simply and directly. He had a good command of English in both speaking and writing.

Bridging Hosts

A brief review of the state and status of the problem up to Stakman's time may help to explain why his work on bridging hosts was so important. Anton deBary established in 1865 that the rust infections on barberries and on cereal grains were different stages of the same fungus. (This actually had been established some years before by a Danish schoolmaster who didn't have the necessary academic credentials and whose work was therefore disregarded.) Ericksson found in the 1890s that stem rust was not simply stem rust, but that there were several more or less distinct "forms" or "biological races" of it (already listed above), each adapted to certain grass hosts. The "more or less" describes the situation that prevailed then, because Eriksson and some later authorities thought that these forms or races were not absolutely fixed but were somewhat plastic or variable and changeable in their parasitic capabilities. The form on wheat, for instance, could be transferred to barley, and after it had been on barley for a time, it developed the ability to infect rye, or even oats, which originally it could not do—the barley served as a bridging host. The rust fungus could also infect many wild grass hosts, and some of these also served as bridging hosts. This concept was the product of extensive and presumably careful research by presumably highly competent and respected men, some of them with the status of major prophets in the field. The same phenomenon was thought to occur in other rusts and also in powdery mildews, which, like the rust fungi, are obligate parasites. For example, concerning bridging hosts in stem rust, Freeman and Johnson stated in 1911 (as quoted by Stakman):[6]

> The barley stem rust enjoys the widest range of any of the biologic forms of the cereal rusts. On the other hand, a transfer of any of the other stem rusts to barley widens the range of that rust. We have here, then, a decided reaction of host upon parasite, enabling the latter to adapt itself to hosts not originally congenial; for instance, W → B → O.

This was important because, if it were so, then attempts to develop varieties of wheat resistant to stem rust (or to other cereal rusts or to powdery mildews) would be hopeless, as some authorities on stem rust already maintained. The stem rust fungus could infect at least 50 species of wild grasses as well as the several cultivated cereals, and so there would be ample hosts to serve as bridges from an old variety to a new one. Or, presumably, the old variety itself could serve as a bridge. Stakman questioned this concept of bridging hosts, even though (or maybe even because) his former preceptor and present research associate and department head, Freeman, espoused it. In the summary of a paper on timothy rust, Stakman and Jensen stated,[7]

> (1) Timothy rust was transferred successfully directly from timothy to *Avena sativa, Hordeum vulgare, Secale cereale, Avena fatua, Avena elatior, Dactylis glomerata, Elymus virginicus, Lolium italicum, Lolium perenne,* and *Bromus tectorum.*
> (2) Attempts to increase the infection capabilities of the rust by the use of bridging hosts for short periods of time were unsuccessful.

Stakman must have been planning for some time a research campaign that would either establish or demolish this concept of bridging hosts. Work on it begun during the research for his Ph.D. thesis was continued and enlarged into a well-conceived, extensive, and meticulously conducted program. Briefly, samples of the different biologic forms of the rust, *P. graminis tritici, P. graminis secalis,* and *P. graminis avenae,* were collected on different wild and cultivated grasses in different parts of the country over a period of several years. Each of these was grown on a set of differential hosts, sometimes for several transfer generations, to separate the components of a mixture, if a mixture of forms was present, or to establish purity of the form if only a single form was present. This was Stakman's idea, one that had not occurred to any of his predecessors, and it went to the root of the problem. Once a collection was purified, it was grown on one or more supposed bridging hosts, such as cultivated or wild barley or a suitable wild grass, then inoculated again onto the main cultivated cereal hosts. Some of the collections, once they were purified, were carried along, with successive transfers every two weeks or so, for four years. Very early in the work, Stakman realized that, first, the original field collections might consist of mixtures of forms that must be separated before any meaningful work on bridging hosts could be done, and second, that the greenhouse work, in which seedling plants were inoculated with rust spores from the many different purified lines, had to be done in such a way as to reduce to a minimum any possibility of contamination between spores of different lines.

He and his collaborators had to design and develop their own facilities and techniques for this, mostly from scratch, and it seems reasonable to suppose that Stakman personally was responsible for much of this, because he already had been deeply involved in it in the work for his Ph.D. thesis. A tremendous amount of rather meticulous drudgery was also involved, as is true of almost any biological research. Supplies of seed of the differential hosts had to be constantly available, including seeds of numerous wild grass hosts; the seed analyst in the Section of Agricultural Botany helped them on this. Each week a couple hundred four-inch pots had to be planted with a dozen to 20 different kinds of seeds, and the identity of the plants in all the pots had to be kept straight. The seedlings, five per pot, had to be inoculated by hand with spores of a given line, with precautions taken to prevent cross contamination by air or persons, and the inoculated seedlings had to be incubated first in moist chambers, then on the greenhouse bench. The plants inoculated with one line had to be kept separate from those inoculated with another. When the rust developed on the seedlings, the infection types had to be "read" and recorded. Stakman gives credit to F. J. Piemeisel, one of his collaborators in this work and his coauthor on several papers. Much of the drudgery must have fallen to Piemeisel's lot, since Stakman was occupied with many other things and also was off on junkets of one kind or another, sometimes for weeks at a time. The results were published in 1918 in a paper[8] that said, in part:

Differential hosts must be used to isolate biologic forms from mixtures before conclusive experiments can be made with bridging hosts.

Puccinia graminis secalis, which does not attack wheat, but does infect barley readily was cultured on barley and other theoretical bridging hosts continuously for three years during which time more than 2,000 wheat plants were inoculated. The rust acquired no new parasitic capability on account of its association with barley.

Puccinia graminis tritici attacks wheat readily, but can attack rye only weakly. Barley is easily attacked. The rust was confined to barley for almost 32 months but it never acquired the power of attacking rye more readily than rust taken directly from wheat.

Barley, which Freeman and Johnson...found to increase the range of parasitism of biologic forms has not been found to do this in the writers' experience.

Several species of Elymus, Agropyron, Hordeum, and Bromus were used as bridging hosts for both the *secalis* and *tritici* forms; but no bridging resulted.

Stakman and his colleagues also passed the different biologic forms of the rust through barberry and found that this did not change their parasitic capabilities.

They made the pertinent comment that if bridging actually occurred, the several biologic forms distinct from one another could hardly have arisen in the first place. This point evidently had not occurred to any of the previous workers on bridging hosts. Concerning the need to use differential hosts to separate mixtures in material collected in the field, they state,

> From the R_1 [rye]—W_1 [wheat] material both biologic forms were isolated. This was puzzling at first, because wheat is not a host for the *secalis* [rye] form. The only plausible explanation seemed to be that the spores of both biologic forms were placed on the wheat during inoculation and not all germinated in the moist chamber. A few viable *secalis* spores therefore remained on the wheat, and when these were transferred to rye, they germinated, causing infection.

It was as simple as that. The concept of bridging hosts was dead. However, they pounded nails into the coffin. Another, somewhat earlier, paper[9] dealt mainly with the new race of wheat stem rust, *P. graminis tritici-compacti*, that Stakman and Piemeisel had just discovered, as described below. In the summary of this paper, they state,

> (9) The facts recorded in this paper, supported by experimental work in the rust nursery and by field observations, indicate that rust resistance is comparable with other permanent characters, and that it is not primarily controlled by seasonal conditions, soil type, geographical location, or other cultural conditions. It is rather an hereditary character, which can not be produced by the accumulation of fluctuating variations within a susceptible line, nor broken down by changes in the host or parasite.
>
> (10) The resistance of wheat varieties may vary in different regions because of the presence of different biologic forms of rust.
>
> (11) There seems to be little basis for the belief that hybrids between resistant and susceptible varieties will exert a harmful effect by increasing the virulence and host range of stemrust.

This work by Stakman and collaborators on bridging hosts was very high-quality research. Basically, what it did was disprove an erroneous but widely held concept, and work that establishes that the major prophets have been preaching false doctrine for 30 years is not usually a highway to fame. If Dr. Freeman was at all bothered by this, he never showed it; in fact, he recognized and frankly admitted that he had gone astray in his own work on the problem. Some of the old-timers must have turned in their graves and some, not yet in their graves, may have undertaken work to check up on Stakman and Piemeisel. If they did, they found that Stakman and Piemeisel were right.

Races of *Puccinia graminis tritici*

At the same time that work on bridging hosts was being done, Stakman was also investigating other aspects of the epidemiology of stem rust of wheat. One of these was the extent to which wild grasses, and cultivated grasses other than wheat, contributed to the development of epidemics of rust on wheat. One of the cultivated grasses that commonly rusted was timothy, and Stakman and Piemeisel gave timothy stem rust a good working over.[10] They collected 11 strains of timothy from different sources, some from as far away as New England, and inoculated them with urediospores of *P. graminis tritici, P. graminis secalis*, and *P. graminis avenae*. No infection resulted from inoculation with spores of stem rust from wheat or rye, but of 3,270 seedlings inoculated with spores of oat stem rust, 57 became infected. That pretty well disposed of timothy and timothy rust as factors in the epidemiology of wheat stem rust.

From there, they went on to rust on other cereals and wild grasses, and their paper on that was a real barnburner.[11] They collected urediospores of *P. graminis tritici, P. graminis secalis, P. graminis avenae*, and *P. graminis phleipratensis* from wild grasses and from cultivated cereals in many different places throughout the range of the rusts in the United States and in portions of Canada, and they inoculated these onto seedlings in the greenhouse. In part, they wanted to determine the role of inoculum from wild grasses in the epidemiology of stem rust on wheat, a very important point. As one illustration of the procedures and results, Diagram I from their paper is here reproduced (Figure 3). Something of the extent and magnitude of the work is indicated by the following statistics.

Urediospores from *Hordeum jubatum* (wild barley) were collected from 49 locations in eight midwestern and western states and from Manitoba, Canada, and used to inoculate wheat, oats, barley, rye, and nine species of wild grasses in five genera. Some of these collections of rust, and some of the other collections, were used also to inoculate barberry, and the resulting aeciospores served as inoculum on several or numerous grass hosts. The paper contains 34 tables, some of them several pages long. For example, Table 28, which summarizes the results of inoculation with urediospores of *P. graminis tritici*, runs to four and a half pages, seven columns per page, 28 entries per column. One of the approximately 125 entries of inoculation results in this table is 4296/4503; that is, of 4,503 plants inoculated in that series of tests, 4,296 became infected (these were plants of *Hordeum vulgare* inoculated with spores from various sources). Several entries in this table give the results of inoculations of more than 100 plants. Tables 30 and 31 each run to three and a half pages. All of this must have involved a tremendous amount of careful and detailed greenhouse work and meticulous record keeping, much of it, or perhaps even most of it, by Piemeisel. At that time, Stakman must also have been teaching plant pathology (organizing and giving the lectures and the laboratory work), and he was also teaching household bacteriology to classes of 90 home economics students. A teaching load like that can be just about a full-time job. He also was organizing and running the new Section of Plant Pathology, shepherding a distinguished visitor on a speaking tour (and serving as interpreter for him), planning barberry eradication, and, perhaps incidentally, courting Louise Jensen, then the mycologist in the Section of Plant Pathology but soon to become his wife. He was young and vigorous then, in the forefront of a new and exciting profession, and in the driver's seat with every intention of staying there. He could rest and sleep later; now there was work to be done. In his oral history, he pays tribute to F. J. Piemeisel as "one of the best graduate students we had had up to that time." When you consider that up to that time they had not had as many as a half dozen graduate students, this may not have been much of a compliment, but Piemeisel must have been good, and he also must have been devoted to work and not given to frivolous pastimes. Piemeisel suffered shellshock in World War I and never recovered from it, a tragic end to a promising career.

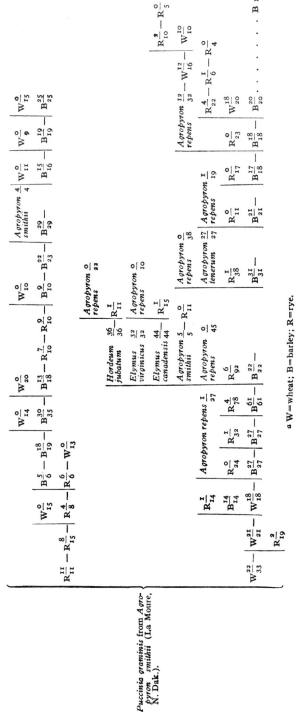

DIAGRAM 1.—Results of successive transfers of rust from wheat and rye in No. 5, Table IV.[a]

[a] W=wheat; B=barley; R=rye.

Figure 3. Diagram from a 1917 paper, showing the careful documentation done by Stakman and his co-workers.

39

The work on rust on wild and cultivated grasses established beyond question that the several distinct forms or biological races recognized up to that time were not uniform, but that each was made up of parasitically distinct strains or subraces. A new and parasitically distinct form or race of *P. graminis tritici* also turned up on wild grasses and on club wheat west of the Rocky Mountains and was named *P. graminis tritici compacti*, from *Triticum compactum*, club wheat. One of the summary statements in the paper[12] states,

> On the basis of parasitism the biologic forms can be divided into two groups:
> (1) *P. graminis tritici, P. graminis tritici compacti*, and *P. graminis secalis*;
> (2) *P. graminis avenae, P. graminis phleipratensis.*

A year later, in 1918, they found this *P. graminis tritici compacti* race to be common on wheat in the South, from Texas to Alabama. If they were making rust surveys in the South then, as they obviously were, then Stakman or Piemeisel or both had already recognized that stem rust epidemics were something more than just a Minnesota or a Midwest phenomenon, or at least that the epidemics might involve something beyond county or state or regional borders. Stakman was already thinking big, and before long he was extending his thinking and his surveys down into Mexico. Who first had the idea of a south-north spread of rust inoculum as an important force in epidemiology is not known, but it probably was Freeman, because in his oral history Stakman mentions that Freeman had him (Stakman) climb up the water tower ladder on the farm campus and expose greased slides to catch spores from the air. The 1918 paper about *P. graminis tritici compacti* in the South says, in the summary:[13]

> The fact that the hard spring wheats commonly grown in the spring wheat region are highly resistant to the biologic form of stem-rust most commonly found on wheat in the south renders less probable the theory of the seasonal spread of rust from south to north in successive waves.

So somebody was thinking of it. Shortly after this, when the new information about wheat varieties and rust races made this south-north spread seem important in the epidemiology of stem rust, Stakman and his co-workers did not have to revise their established thought patterns to take it into account; they already had done most of the necessary thinking and developed most of the necessary techniques to effectively study this aspect of rust epidemiology, one that turned out to be very important indeed.

In his oral history, Stakman gives some interesting sidelights on the work that led to the discovery of the new race of wheat stem rust, *P. graminis tritici compacti*. A terrific epidemic of stem rust had practically wiped out the wheat crop in most of the main spring wheat states of the midwestern United States and the prairie provinces of Canada. M. A. Carleton was then head of the Office of Cereal Crops and Diseases of the USDA, and Stakman, who was a collaborator in that office, braced Carleton for expense money to collect rust samples in the Pacific Northwest, beyond the Rocky Mountains.

> So I asked him for this permission and I told him that it was urgent because I wanted to study the rusts and make collections on the way out to the Pacific Coast, and the weather had turned hot and a lot of the wheat was drying up fast. So he sent back a telegram promptly and said, "Authorized to go as far as Great Falls, Montana," and I telegraphed back, collect I hope,

and said, "Important to go west of Continental Divide." He sent back a telegram and said, "Authorized to go as far as Great Falls, Montana," and that seemed to settle the matter because I knew him pretty well and we were getting to be pretty good friends, but he was a stubborn Dutchman (at least he once told me that he had a little Dutch in him). And I didn't think he'd change his mind very easily and I had to hurry, so I went downtown and roamed around the railroad ticket offices[14] and tried to find out what the difference was between the cost of going to the West Coast and of going to Great Falls, Montana.

I was planning to pay for the difference myself if I had to, but luckily I found that it was cheaper to go out to the West Coast and back, as far as railroad fare was concerned, than it was to go to Great Falls, Montana, and back, because there were summer excursion fares that covered the West Coast but didn't include Great Falls, Montana. So, knowing that Carleton was a bargain hunter, I sent him a telegram and said, "Cheaper to go to West Coast than to go to Great Falls, Montana." And he sent back a telegram and virtually said, "For heaven's sake, go ahead and quit bothering me." I think he just said, "Go to West Coast," but you could just read in the telegram that he said, "Get out of my hair and go ahead."

Well, at any rate, I went out and traveled by train, went out on the Great Northern, went through some of the best wheat-growing areas of Minnesota. Actually I started way down in southeastern Minnesota and then went out through the heavy wheat-growing areas of North Dakota, eastern Montana and western Montana and on out into Washington and so on.

For a thousand miles from eastern Wisconsin out to the high plains, almost to the mountains in Montana, I didn't see any wheat fields that would yield more than six or seven bushels an acre of chicken feed. It was ruined. Everything was ruined all the way along the line, and some of the grain was still somewhat green. Well, I traveled by local trains and stopped at strategic places and deployed out from there. But I used to ride up in the smoking car, not because I wanted to smoke, particularly, but because that's where the brakeman was when he wasn't somewhere else, and that's where he'd stay, if there was a long way between stations, and I just had a pact with him that he would watch me while I ran away from the train into nearby fields and made collections. I told him what I was doing, and asked him if he'd give me the high sign, as they call it in railroad parlance, when the train was about to start, and he did, and so I never missed a train. Otherwise I might have missed several.

Well, every place the train stopped I ran out, if there were grasses or grains nearby, and made collections, made a notation of the place and whatever else I could. Some of these brakemen got interested and they helped out on the thing quite a little. Well, I sent these collections of the rust back to Piemeisel in St. Paul, and he made inoculations with them to identify them. I had told Carleton in the original letter that I wanted to get west of the Continental Divide because I felt that if there were different races, the best possibility of finding them quickly would be to get into an area other than that in which we had been working mostly. And at the first station beyond the Continental Divide, I jumped off the train and went out and grabbed a handful of rusted wild barley and sent that back, and the very first collection that I got west of the Continental Divide turned out to be something different, which I learned by telegram from Piemeisel when I got out to the West Coast.

He continued collecting samples of rusted grains and grasses through Idaho, Washington, and Oregon and returned via the Northern Pacific.

What we found was, when we started working on this, that virtually all the collections that we got from that area, west of the Continental Divide, were different, very sharply different, from the eastern collections in this respect: There were half a dozen spring bread wheats that we used to inoculate because they were all susceptible; one of them was Marquis. Another was Haynes bluestem, or just Haynes, and Glyndon fife—these three particularly. Now this first collection that I got in the West didn't attack these varieties normally. Instead of producing big fat pustules, it produced smaller pustules with a sort of halo around them, which we called a type two infection. This proved to be true of other western collections also. Otherwise, as far as we could tell, they were like eastern collections. If they differed on one variety, and if this new rust that we got didn't hit the varieties that had been susceptible all the time, the question was, would some collections of it hit something that had been resistant up to then? I mean, you simply worked both ways. So we tried this out on some of the durums and various others, and pretty soon we tried a variety known as Kanred, which had been resistant. We inoculated that with a number of collections, and instead of being resistant, it was susceptible to some, and so we had a new strain of rust that couldn't attack the varieties that had been susceptible but could attack a variety that had been resistant in the Midwest up to that time.

Well, of course, then you just simply selected varieties from different species groups of wheat—that is, the bread wheats, the macaroni wheats or durum wheats, and the Polish wheats, the wheat species groups . . . I suppose we inoculated at least 250 varieties from these various groups with the rust collections and finally worked out a key for the identification of the races. And after a long time we simply found out that there are an indefinite number of races [he is quite a ways into the future, now]—there isn't any question about it. I mean, there is just an infinite variety of races of wheat stem rust. Also that gave us the idea—we were finding similar evidence for various other fungi at the same time, the flax wilt organism, for example, and *Helminthosporium*—and that led to a project on the identity and origin of parasitic races or physiologic races of plant pathogens in general.

Throughout much of his career as head of the Section and, later, Department of Plant Pathology, Stakman slanted much of the research of nearly all graduate students toward variation in fungi, and particularly phytopathogenic fungi, those that cause diseases of plants, principally economic plants. That is, over a period of 40 years, his work and that of his students (up until very near his retirement, all graduate students in plant pathology at the University of Minnesota were Stakman's students, in that he, not the staff member with whom the student happened to be working, was the student's principal advisor) was on the nature, cause, and extent of variation in fungi. Literally hundreds of papers came out of the department on this subject. Stakman thought, talked, wrote, lectured, preached, and propagandized on it endlessly. All of this went a long way to establish the concept that fungi share with the morphologically more complex organisms like higher plants and animals the ingredients of evolution. But it all started with work on rust races (Figure 4).

A third biologic form or race of stem rust of wheat was described by Levine and Stakman in 1918.[15] It was collected by Piemeisel on volunteer wheat near Stillwater, Oklahoma, on October 18, 1917. "About a dozen" forms or races of *P. graminis tritici* had been identified by October 1918.[16] The significance of these races in the development of new, rust-resistant varieties was pointed out in 1921.[17] Keys for the identification of races, and a table giving the reaction or infection type of 37 races of *P. graminis tritici* on the standard 12 differential varieties of wheat were published in 1922.[18]

That the system developed by Stakman and his many collaborators and co-workers for the identification of races of wheat stem rust was sound is indicated by the fact that it soon became used throughout the world where research on wheat stem rust was being done.

The procedure is still valid today and continues to be used, mainly to identify the races present in a given region and enable the plant breeders to keep ahead of them. It is occasionally modified by including varieties in addition to the 12 standard differentials to identify additional races or subraces, such as 15B and 15B-1 and so on, that have or might become important here or there. But the scheme and techniques, and even the varieties of wheat used for the differentials, remain basically the same as when Stakman and his co-workers developed them in 1915–1920. Hundreds of races have been identified, and thousands could be if there were any point in it. Each year since the early 1930s, an annual survey has been taken of physiologic races present in the United States and a record kept of the percentage of the total collections that each of the principal races makes up. This is an integral part of the ongoing study of the epidemiology of stem rust that Stakman pioneered. Another of his contributions was the development of the program for the eradication of barberries, the alternate host of the stem rust

Figure 4. Stakman in the rust laboratory, August 1922.

fungus, but before that is taken up, a look at some of Stakman's nonresearch activities of this period seems appropriate.

Travels—1913–1915

Stakman wanted to travel, for knowledge, not sight-seeing.

But there were a lot of things that I still wanted to do, and I wanted to travel more. During these semi-pro years, . . . I did quite a little writing for one or two magazines. There was one magazine that had an editor, fine old man, who sort of liked the kind of English that I wrote at that time; I don't write that kind of English any more. He liked the essay style, the old fashioned essay style where you start with Adam and Eve and you build it up until eternity and beyond. He liked that, I don't know why, but he did. He was *Bulletin* editor in the Department of Agriculture at Minnesota for a while, and then he ran an agricultural paper, and he told me that he'd take all the stuff that I would write and pay me a cent a word for it.

Well, actually, in those days that wasn't too bad, so I set aside a travel fund and put this money into it, and then put various other sources of income into it and always built up a travel fund.

I want to say right now that in those early days, when the salaries were low and everything, and when you were encouraged to do this sort of thing and take pay for it, I did. But after the salaries became adequate I refused to take money for writing or for talking. I want to make that clear, because I think that when a person is on the staff of a state university and he is on full time and he's getting an adequate salary, he owes the university full time, and if he takes too much time away from the university, it seems to me that he isn't—well, I question the ethics of it.

What Stakman considered to be an "adequate" salary for his staff members and what they considered to be an adequate salary differed considerably; some of his staff members were inclined to believe that he favored an arrangement in which the staff members would work for 12–16 hours a day, 365 days of the year, with no vacations or time off, preferably at the lower salary range for whatever rank they had, and with no outside income. As a matter of fact, some of them did so. Granted, throughout most of his tenure as Section or Department Head, salaries were pretty low everywhere—but in Plant Pathology they were lower. There were some compensations, however.

Well, all right, I wanted to travel. The first long trip which could be considered an educational trip was in late 1913 and early 1914, when three of us set out to attend the meetings of the American Association for the Advancement of Science in Atlanta. We went there by way of Illinois, Ohio, Virginia, North Carolina, South Carolina, Florida, and visited a lot of educational institutions on the way and learned everything that we could, and then returned by way of Tennessee and the Missouri Botanic Gardens at St. Louis and the Washington University, where they had some very good work in botany.

I had hoped to travel to Europe as soon as possible and spend a little time with Tubeuf at Munich. He was a famous pathologist at that time. But Europe came to the United States, more or less, in the summer of 1914, in the form of Dr. Otto Appel. Dr. Appel was one of the outstanding plant pathologists in the world at that time, and he was considered probably the best authority on potato diseases, certainly in Europe and maybe in the world. He was invited to the United States to study potato diseases and potato production in this country and make recommendations as to how the potato industry could be saved, because diseases were accumulating.

Fortunately Minnesota was a potato-producing state, and I was authorized to go on this trip. Dr. W. A. Orton, [who] was the head of the party, and Dr. William Stuart, who had charge of all research on potatoes in the United States Department of Agriculture, and a number of other specialists remained in the party all of the time. I was authorized to go, as I said, by the University of Minnesota.

We went through Nebraska, various places, Colorado, and Utah, Idaho, Washington, spent more than a month, five or six weeks I think it was, and . . . I had an opportunity to associate with these men who were among the best in the world on potato problems, and also to meet a great many specialists in the states, and to meet some of the biggest potato growers in the United States and to talk with them and find out what their problems were.

But the greatest privilege I had on that trip was acting as *Dolmetscher* for Dr. Appel—that means interpreter. He understood some English but he didn't speak it, and he made speeches, in German, wherever we stopped. This was sort of a traveling Chautauqua, and almost a traveling circus, and so I translated Dr. Appel's speeches for the crowd.

At the beginning he would talk until he got a little windblown and then he'd look at me and I'd try to translate word for word, and at the end of that time—I remember Orton used to look sometimes a little puzzled and sort of mutter to himself, "Are you sure you translated that right?" (Orton had been in Germany and knew some Yankee German.)

Appel got the sign language and he said, "ja." But later on he said to me, "I think that you can do just as well if I go on and talk and you paraphrase my talk. You just give it in your own words," which is what I did, and so I learned all that Dr. Appel knew, in a sense, because I think he talked so often that he must have told almost all that he knew, and he must have told all that he knew several times over because he made many speeches on the same subject.

Well, okay, I learned a lot from him. I also got a lot of confidence in the sorts of things that I thought I knew, and I got a lot of ideas from him as to what plant pathology was like in Europe, Germany in particular, and got the idea of state intervention and participation in campaigns for the control of diseases, because they did a great deal of that over there. They had a potato certification system, for example. So I learned something about, shall we say, pathological but scientific statesmanship from him, the relation between science and the state, because Germany at that time was still a monarchy.

Well, the war broke out in August, you remember, of 1914, and Appel had to stay in this country for quite a long time before they'd let him go back, and the result was that I saw him occasionally and got still better acquainted with him. But that was as good as going over and studying with Tubeuf, probably a lot better. Then at the same time Dr. Johanna Westerdijk, who was in charge of the Willie Commelin Scholten Phytopathological Laboratory in Holland came over. She and Appel were good professional friends and I got well acquainted with her. Then the next year the eminent Danish pathologist, Koplin Ravan, came over and I made another long trip with him, and so I felt that I had gotten quite a little experience out of these scientific travels, because I had done it with eminent men from this country and from Europe.

The oral history contains more about Appel:

It was particularly agreeable to be associated with him because he was a very fine man. He was a tall, upstanding, brown-eyed and brown-haired, very kindly individual. [He had a nice smile and] also a scar that went clear across his face from dueling because he had gone to the university when the "Code Duello" was still in existence in the university. He always said that he shouldn't have had that scar but it was the fault of the referee, that he ... had called "halt" because he'd wounded the other fellow. I think this was in the Code, that if anybody said stop, they'd stop. Well, the referee is supposed to hit their swords up, because he stands here, and he's supposed to intervene. Well, he said this referee hadn't done it. Otherwise he (Appel) would have come from his duel without a scratch. This may be true because I think he had a long reach, and I watched his wrist and it looked as if he could still probably do a pretty good job of fencing.

[About Miss Westerdijk:] Well, I think that Johanna Westerdijk and I became friends also. She didn't go on this whole trip. She met us I think out in Washington. But they (Westerdijk and Appel) came to St. Paul later, and in a sense I guess I was their guide there and informal host. Dr. Westerdijk talked English perfectly well and so she did quite a little of the translating for Appel. But we went out on field trips together (Figure 5) and I got

Figure 5. A field trip in 1914. Left to right: E. M. Freeman, Johanna Westerdijk, C. O. Rosendahl, E. Louise Jensen, E. C. Stakman, and A. G. Tolaas.

acquainted with her later. I'll tell you what kind of a woman she was. At that time, if women smoked cigarettes it was considered a badge of [their] trade, and Dr. Westerdijk smoked cigarettes. Women started smoking cigarettes in Europe earlier than they did here. I saw girls, for example, smoking cigars in Denmark, in Copenhagen, when a few girls were just barely starting to smoke cigarettes over here. We were at a restaurant in St. Paul, and this was supposed to be a very advanced restaurant, you see. They served drinks. They did a lot of things that they didn't do in the ordinary restaurant. But just for fun, I offered Dr. Westerdijk a cigarette. I think Dr. Appel was an inveterate pipe smoker and also smoked cigars. We'd been smoking our pipes or something so I asked Dr. Westerdijk, I said, "Would you like a cigarette?"

She said, "Yes, I would like one," and I sort of shrank back; I thought, oh-oh, what's going to happen? So I offered her a cigarette and offered her a light and she lit it and smoked it, and the waiter came over. He said, "I'm sorry but women are not permitted to smoke here." "Why not?" "I don't know, I guess a rule of the house, something like that." She said, "Why, I thought this was the land of the free. I smoke cigarettes at home, why can't I smoke them here?" Of course all the poor waiter could do was say, "I don't know, I'll go and ask." "Well," she said, "you go and ask your superior." This went on, back and forth, for some little time, but Miss Westerdijk finally finished her cigarette, and they were served dinner.

CHAPTER SIX

The Barberry Eradication Campaign

That barberries (*Berberis vulgaris*) had something to do with the development of local epidemics of stem rust of cultivated cereals was recognized long before the connection was experimentally established in 1865. Laws requiring the eradication of barberry bushes were passed in France in the 1600s and in New England in the 1700s. However, the role of barberries in the development of the rust on cereals was by no means universally agreed upon, even among plant scientists as late as 1915–1916 and even in the region that had just suffered a devastating epidemic of rust on wheat. Stakman and Piemeisel, in their 1917 paper already cited,[1] state that

> Pritchard . . . as a result of experiments in North Dakota, concludes that barberry plants have but little effect in starting rust epidemics in the spring because—"*P. graminis* probably appeared upon the experimental plot of winter wheat almost or quite as early as upon *Agropyron repens* and *Hordeum jubatum* even when the latter were in the immediate vicinity of the barberry. . . . one form of *P. graminis* is common to *Hordeum jubatum*, *Agropyron tenerum, A. repens, Avena fatua*, oats and rye, but is incapable of infecting either barley or wheat. This furnishes little encouragement to those who believe that *P. graminis* is spread to the wheat fields from the barberry bushes or from occasional protected spots, as beneath ice, by the aid of the native grasses."

I do not know what Pritchard's qualifications were for making statements such as those about the biology of the stem rust fungus, but the statements were from a paper published in the *Botanical Gazette*, a respectable scientific journal, so he must have been a member in good standing of the scientific brotherhood, and both he and the editor of the journal must have thought he knew what he was writing about. He didn't. The quoted statements make absolutely no sense now, and they could not have made any sense then. But if a professional scoffed at the idea that eradication of barberries would reduce rust infection on cereal plants, proponents of the idea of compulsory and complete eradication of barberries might expect to encounter some opposition among the common people, especially those who had prize barberry hedges on their lawns, or nurserymen who grew and sold barberry bushes, or those sturdy individualists who resent anything compulsory, especially from government authority of any kind. The wheat farmers were not likely to be opposed to eradication of barberries, but not all farmers grew wheat, and some of those who did not saw no reason why they should seek out and dig up barberry bushes on their property or pay someone else to do so. Stakman knew that some convincing would have to be done, and he met the challenge head on. As soon as he was put in charge of the federal barberry eradication program, he mounted a propaganda campaign that would have done credit to any promoter or public relations expert of modern times. Due in great part to Stakman's personal drive, the campaign became to a considerable extent a crusade, and by the time he retired as director of the program 15 months later, it was so well established that its continuation was assured. The campaign was very fortunate in timing because it got under way in the spring of 1918, shortly after the United States had entered World War I, the war to make the world safe for democracy, and one of the slogans was "Food will win the war." Food meant wheat. And with the terrible stem rust epidemic of 1916 fresh in everyone's mind, wheat production meant control of stem rust. Control of stem rust meant eradication of barberries, or so it could be argued, and Stakman argued it powerfully and in high places. Among the war emergency boards established in 1917 was one of plant pathologists, on which Stakman held an important post as Commissioner of Research. Eradication of barberries wasn't really research, but Stakman disregarded that technicality and got the campaign under way. He tells how it started, in early 1917, with a meeting at Fargo, North Dakota.

Henry Luke Bolley was the plant pathologist and state botanist at the North Dakota Agricultural College at Fargo. He was a real pioneer and he also had tried to develop some rust-resistant varieties. He suggested to the people that were organizing a Tri-State Grain and Stock Growers' Convention that they invite me to come up and talk on rust and its control, because we'd been publishing on it. I knew Bolley very well and Bolley was a Hoosier and he was keen enough to know that if you invite somebody from outside of the state to come in, maybe they'd listen to him more than they would to people within the state whom they'd known for a long time. That's the same old story, a prophet is without honor in his own home or something like that.

At any rate, I accepted the invitation, and it happened that at that particular time Dr. A. H. R. Buller, who was professor of botany at the University of Manitoba, in Canada, was visiting with us at Minnesota, and I suggested to him that he stop off with me on his way back to Winnipeg, at this meeting at Fargo. He was a botanist, pure botanist, academic botanist, not a plant pathologist, but they didn't have plant pathologists up in Canada, at least not many of them. And M. A. Carleton was there because since this was an interstate convention, the head of the cereal work of the federal Department of Agriculture should be there.

So Bolley and Carleton and Buller and I had a strategy session before the meeting, which was in January 1917, and decided more or less on what I should emphasize. So in my talk I think I emphasized the fact that more money effort should be put into breeding for resistance, on the basis of what we already knew. The second thing, that this group who were taxpayers and had a lot of influence on legislation, they should demand that the federal government—and recommend that the state governments—find out more about the epidemiology of the rust: Where did it come from? Under what conditions did it develop most abundantly? In other words, that there should be this research program. The third one was that they should recommend at least the local eradication of barberries because it was known that a great deal of the rust came from barberries. Whether all of it came [from there] we didn't know, but they should do these things: First, push the breeding work and supplement that with research to get it on a sounder basis; second, study the epidemiology of rust throughout the country and beyond if necessary,[2] and only the federal government could do that, obviously; and the third one was to recommend the eradication of barberries, at least on a local basis.

Stakman was on fairly solid ground concerning the recommendation to eradicate barberries, at least locally, because he and his co-workers had been making detailed observations on rust development near barberries in Minnesota since 1909, when he began his graduate work. Stakman must have made one of his usual powerful presentations, because after the meetings, Bolley and Carleton were asked to draw up a set of resolutions, which they did. The resolutions were passed immediately, and

> the consequences of that meeting were far-reaching, because they passed a barberry eradication law in North Dakota that very winter, in 1917, and that was one of the direct results of this meeting and of the agitation that preceded it and followed it. Well, okay, so there was one state in the United States then that had a barberry eradication law; that was North Dakota.

Another result of the meeting was that the USDA decided, most likely largely as a result of the investigations by Stakman and Piemeisel and at Stakman's urging, to undertake more extensive work on rust epidemiology, especially on the identification and distribution of races of *P. graminis tritici* and on interregional spread of the rust from south to north, both of which aspects were pursued with zeal for the next 20 years or more by Stakman and his collaborators and students.

Very soon after America joined the war in 1917, various war emergency boards were appointed, among them a War Emergency Board of Plant Pathology, with six members, all outstanding plant pathologists. H .H. Whetzel, head of Plant Pathology at Cornell University, was chairman of the board, and each member was a commissioner of something or other, Stakman being Commissioner of Research. The commissioners toured the country drumming up support for plant disease control as a means of increasing food production. They helped make people conscious of plant diseases, of plant pathology, and of at least some plant pathologists, and they helped bring plant pathology and plant pathologists out of the laboratory and into the fields to become more a part of food and fiber production than they had been before. During these travels, mostly by train, Stakman had ample time to convince other members of the commission that control of stem rust of wheat was a priority project, that barberry eradication was an essential part of that control, and that the best way to go about this would be to mount a national barberry eradication campaign, with state laws to make the eradication of barberry bushes mandatory. Plant pathologists in the central and midwestern wheat-growing states discussed this idea pro and con. Bolley, a plunger by nature, was all for it, but for some plant pathologists, compulsory eradication was pretty strong medicine, and they shied away from it. Stakman says,

> The obvious fear of rust really made it possible for the bolder people to push hard now for a national barberry eradication campaign. Heretofore

they'd been advocating voluntary eradication, but with the food shortage such as it was and the threat of a disaster from rust, the bold people got bolder and some of the more timid ones got a little less timid, perhaps.... Many of them started a "stop Bolley" movement because they were afraid that this agitation would come to their states and they weren't ready to advocate a law to compel the eradication of barberry. They hadn't objected particularly to voluntary eradication, but when it came to ... an education or research institution advocating a compulsory law, that was a new precedent, going a little bit too far.

A strategy meeting was held in Chicago early in 1918, attended by plant pathologists from the central and midwestern wheat-growing states, and those in favor of the national barberry eradication campaign prevailed. As a result, a delegation consisting of Stakman, Bolley, and one other person (who was not identified and who had little role in subsequent events) set out for Washington to approach people in the federal government regarding support for a campaign of barberry eradication, to begin at once.

We left for Washington in the afternoon, from Chicago, right from the meeting. Bolley kept talking to me in the smoking compartment of the Pullman car until 3 o'clock in the morning, and I think he was trying to indoctrinate me. At the end of that time he said, "You know a lot about rust, and you've got a considerable amount of vigor, and it would be a good thing if you would channel that vigor in the right direction, and this time it's barberry eradication, rust control."

The next morning they met with people in the Bureau of Plant Industry, with whom, of course, they had been in regular contact, since much of Stakman's work on rust up to that time and, indeed, for many years later, was supported in part by USDA funds. Also, Stakman was a firm believer in going through channels. They then went to put their proposition up to the Secretary of Agriculture, David Houston. Houston had already been briefed by congressmen from Minnesota and the Dakotas, but he could hardly have been educated in depth on the technical aspects of stem rust and its control. Stakman and Bolley made their pitch and, when they had finished, the secretary simply said, "I'll do all I can to promote it."

And he had some funds that he allotted for barberry eradication immediately, instead of waiting for additional appropriations from Congress.... At that time Bolley said, "Fine, and you put this man in charge," and he hit me in the stomach, almost knocked the wind out of me, and this is what happened.

The amount of money made available immediately by the Secretary of Agriculture was not stated, but later Stakman says that for the first two years the budget was $150,000 per year, so he must have received at least a moderate fraction of that right off. Even if he received only $25–50,000, that would be equivalent to a quarter to a half million dollars in 1981 dollars. And just on the basis of a brief oral appeal and a smart slap in the stomach by Bolley. It isn't done that way any more. (Red tape was already alive and well then—Stakman later tells of putting in an expense account during his travels for the USDA in Alaska, in 1919, that included a 10-cent tip for dinner; he received a query from the disbursing office requesting to know whether this 10-cent tip was for one person or two.) But even given the fact that those were somewhat simpler days, this whole deal has about it an aura of unreality. Two men essentially unknown to him approached the Secretary of Agriculture with a proposal to eradicate all the barberry plants throughout an area equal to about one fourth of the United States; no one knew if it could be done; nothing on that scale had ever been tried before, anywhere. Even among supposedly knowledgeable plant pathologists, Stakman and Bolley had had to twist some arms to get backing for this approach, and even Stakman's boss, Freeman, was a little hesitant and dubious. However, when Stakman was convinced of something or of some course of action, he could be very confident and persuasive. He convinced the Secretary of Agriculture, and he left Washington with the authority and money to start a campaign to eradicate all the barberries in 13 states from and including Ohio and Michigan on the east to and including Montana, Wyoming, and Colorado on the west. By the early 1940s, when barberry eradication was merged into the Plant Pest Control Division of the Agricultural Research Administration, nearly 300 million barberry bushes had been destroyed on 126,000 different properties. Probably no one, not even Stakman, had envisioned a task of that magnitude. By 1954, more than 450 million barberry bushes had been destroyed in 18 states.

So I went back home and had many sleepless nights, because we didn't know enough to tell exactly how much good barberry eradication could do.... I spent the next 15 months, starting in February, 1918, propagandizing for barberry eradication and organizing the barberry eradication campaign. The first strategy session that we had was with a Minnesota plant pathology seminar in February on a very cold night when I invited the seminar down to our home and told them what was happening and said, "Now, how shall we run the campaign?"...the next most important thing to do was to make sure that you could organize federal-state cooperation, because this had to be a regional campaign, interstate, and at that time the relations between some of the states and the federal Department of Agriculture were not too good. The states sometimes thought that the U.S. Department of Agriculture infringed too much on their preserves, and some of them thought the federal department didn't help them enough, so sometimes they were a little tense and sometimes there was friction between them. In any case, this was a campaign in which the legal requirement, if there was to be a legal requirement [i.e., laws making eradication of barberries compulsory] had to be a state requirement, not a federal requirement.

There were also jurisdictional difficulties and disputes within some of the states that had to be ironed out or smoothed over. Stakman also wanted to bring industry and business into the campaign, to take advantage of their financial resources, administrative expertise, and drive, whereas others were skittish about this, preferring to maintain their academic aloofness, although the academic hand had always been held out discreetly from beneath the academic robes to receive whatever largess industry might drop into it while the academic face was averted. Farmers, industrialists, and academic researchers were not then members of a team striving for a common goal as, at least to some extent, they later became. Stakman, however, was not about to let any theoretical objections stand in the way of having industry take part.

Then there was also the problem of overcoming a certain distrust that many scientists had of industry and of big business. One reason perhaps was the difference between town and gown, and many of the scientists were more or less academicians, and they figured that they were working for all the people and not just for big profit-making oganizations. And so there was a certain suspicion and a sort of aloofness between them. . . . Obviously it seemed that if this campaign was to succeed, you had to have influential people to push it, and insofar as scientists, shall we say, are usually poor lobbyists, and are suspected by money-granting institutions of just wanting to build their institutions for their own satisfaction, it seemed best to have the people who were really affected by rust, to have them help assume the responsibility for carrying things through.

Some of the leaders in the grain milling and merchandising industries, the railroads, and other agriculture-related enterprises were looking for an opportunity to contribute. Among these was F. M. Crosby.

One of the leaders, I'd say the leader really who started cooperation, was Mr. Franklin M. Crosby, who was vice president of General Mills, in charge of their wheat department, and he was quite a scholarly individual and also a very good businessman. He asked whether the miller or the grain men could be of any value, be of any help, in selecting rust-resistant varieties. Well, I had to say that this method had been tried, but with poor results. And in the meantime he had been reading about rust. He was a Yale man and so he had been reading President Dwight's diary, and he encountered a barberry story in there, more or less what was known about it then, and then he got somebody to translate the Danish barberry eradication law. He proposed that there be a meeting to discuss this. We'd talked about rust and he said he thought this should be more widely known. So he organized a meeting at the Minneapolis Club,[3] and the governor of the state was there and some of the state officials and many members of the grain trade and milling interests, the railroads, bankers, wholesalers, farm machinery people, and the business community generally.

Stakman presented to them the case for rust control and barberry eradication, and he must have done it well because

at the end of the speech they seemed to be very interested, as they should [have been] because this all affected the business and prosperity of that area. And so as an aftermath of this, they suggested that possibly they could help by establishing an organization, which I think at first they called The Rust Prevention Association [and which later became the Conference for the Prevention of Grain Rust. As the Crop Quality Council, the organization still is very much in business]. They did three things: They helped push the barberry eradication campaign; they also helped support the breeding work; and they helped solicit funds for additional research work. I would say that one of the two greatest accomplishments conceptually in this barberry eradication campaign was really helping to improve the state-federal relations in agriculture, and certainly [the second was] in helping to improve the relations between business and science, and without the help of this organization of businessmen I don't believe that the barberry eradication campaign would have been permanently successful.

Later he again pays tribute to some of these men.

Some of these businessmen were real men of vision, and I want to pay particular tribute to the Crosbys, and particularly to Franklin M. Crosby.... His older brother, John Crosby, was also a man of vision.... one of the other men that it seemed to me had remarkable vision was Ralph Budd, who was then president of the Great Northern Railroad, later became president of the Burlington.... Another very fine one was Charles Donnelly, president of the Northern Pacific Railroad. He wasn't quite as active in a sense as Budd, but I'd say that Franklin Crosby and Ralph Budd were the people that you could always go to and they could get things done. They were perfectly willing to devote a lot of time to this, and they always seemed to have time. Very efficient. Another one was C. C. Weber, who was president of the Deere-Weber Farm Machinery Company.... They constituted some of the officers, or they were among the pillars of strength, in this organization that was formed and helped get funds to support the organization. They got contributions from the business community to help. They took over certainly not only a lot of the educational work, the propaganda, but also they cooperated very effectively in getting appropriations for the continuance of the campaign and for the research work.

Eradication of barberries began immediately in 1918 with Stakman in charge of the program. An office staff was established, including a professional writer to prepare articles for newspapers and all kinds of farm and agriculture magazines; field staffs were organized in the different states; Stakman himself coordinated all efforts and traveled far and wide making speeches to all who would listen and who might contribute in any way to the program. Exhibits were prepared for schools and for county and state fairs—not a single exhibit, but a multitude of exhibits. Some of them illustrated the barberry bush, including mounted leaf clusters; others were posters of various sizes showing the life cycle of the stem rust fungus. These went to every school—and to nearly every schoolroom, country, village, or city—in the 13 states. Similar ones went to every post office and to farm implement dealers, feed mills, grain elevators, and railroad depots. A tremendous propaganda campaign ("Free The 13 States From The Menace Of The Red Tyrant!") was undertaken. Boy and Girl Scouts, women's groups, county agricultural agents, farm organizations, garden clubs, all joined in. Even the nurserymen's associations joined. (Some nurserymen were reimbursed for the financial loss involved in destroying their barberry bushes; some were not.) By the mid-1920s, if there was anyone in the 13 states over the age of six who was not aware that barberries were an enemy of mankind and must be destroyed, that person must have been dull indeed.

The Conference for the Prevention of Grain Rust aided materially in this but did not become formally organized until the fall of 1922, when it had its second annual meeting, held in the Plant Pathology building on the University Farm campus in St. Paul. The roster of people attending and the progress made in barberry eradication up to that time are shown in the proceedings of this meeting (Figures 6–9).

Donald G. Fletcher was introduced at the meeting. He had been borrowed from the USDA to become Executive Secretary of the Conference for the Prevention of Grain Rust, and he stayed on in that capacity or as Executive Vice President, through the several changes in name and function of the organization, until his death in 1968. By then the organization had for some time been the Crop Quality Council, concerned with promoting interests of agriculture and farm-related industry. Fletcher was a very effective lobbyist with industry and with state and federal legislators and agricultural organizations; he was suave, urbane, and knowledgeable. He left a fine record of achievement for the general good of agriculture.

Stakman also promoted the cause by writing technical and popular articles on stem rust and barberries and barberry eradication. The 1918 USDA Yearbook included an article by him, "The black stem rust and the barberry," which was reprinted as a separate publication.[4] Among his popular articles was one, "Black Rust of Wheat, Caused by the Barbarian Barberry Bush,"[5] that included a somewhat corny cartoon showing the black stem rust in the form of an Old Man of the Sea riding on the back of a faltering bundle of wheat; it was entitled "The Wheat Man's Burden." Another was a USDA bulletin, *Destroy the Common Barberry.*[6]

Stakman returned to the university in May 1919, and Dr. F. E. Kempton, of the USDA, was put in charge of barberry eradication. Stakman continued to work on rust and continued as an agent of the USDA. He traveled widely in the United States, Canada, Alaska, and Mexico, and as far south as Honduras. Early in 1922 he went to Europe, at the request of the USDA, to evaluate the effects of barberry eradication in various countries there. Mrs. Stakman accompanied him. They visited Wales, England, and Scotland, then went to France, Spain, Italy, Greece, Yugoslavia, Austria-Hungary, Germany, Denmark, Sweden, and Norway. As he was moving his baggage out of their second-class compartment at a transfer point in Turin, Italy, he was bumped and jostled and had his wallet lifted by an expert pickpocket. He had secreted most of his cash in other places about his person and so did not lose too much money, but all of his expense records and receipts were in the stolen wallet, which was disturbing because the USDA disbursing office required receipts for claimed expenditures. However, the wallet, with all the expense records, later turned up at the American Express office in London; the thieves, after taking the money, had dropped the wallet in a postbox. He and Mrs. Stakman sat up all night on the train from Greece to Yugoslavia, although many berths were empty—Stakman was not aware of the custom prevailing in that part of Europe requiring that one tip both the travel agent who made the arrangements and the wagon-lit's conductor in order to secure a berth.

He accumulated plenty of evidence to show that where barberries had been eradicated, the rust had been eliminated. All the evidence in the many different countries agreed on this, with absolutely no room for hedging or questioning, and after Stakman returned he stated this, forcefully and repeatedly, in his talks and writings. This evidence was important for the continuation of the barberry eradication program, and, partly as a result of his presentations, the program was continued with even greater support. Up until this time, there had been some opposition to the program—it was a boondoggle and a waste of the taxpayers' money; it was getting too much of the money available for agriculture; it could not possibly succeed; it was not being run the way it should be. Any program of that magnitude was bound to encounter some opposition. Stakman had anticipated this and was prepared to meet it, which, of course, was a major reason for the European tour, which he most likely was responsible for promoting. Any opposition based on an honest opinion that eradication of barberries would not effectively reduce local epidemics of stem rust was quieted by the information gathered from the European trip and publicized so effectively by Stakman after his return. Much of this information he summarized in an article entitled "Barberry Eradication Prevents Black Rust in Western Europe."[7] In this, he described the situation in various countries under headings such as, "How English Farmers Prevented Rust," "Barberry and Black Rust in Wales," "Rust in the British Isles Depends on Barberry," "Denmark Kills Barberry and Stops Rust" (this included an account of peasants, armed with axes, invading an estate to destroy the barberry bushes that were responsible for the rust that destroyed their grain), "No Black Rust Without Barberry in Norway and Sweden," "Little Black Rust in Central Europe," and "Europe Continues the Fight" (this was to get across the idea that eradication was not just a one-shot remedy but required continuing effort). In passing, he quoted from an 18th-century French agricultural treatise, Duhamel's *Traité des Arbres*, the author of which objected to eradication of barberries on the basis that, "Many insist that the flowers of the barberry [the author so designated the yellow clusters of aecia on the barberry leaves] prevent wheat from producing kernels. I have not investigated the matter but consider it improbable." At least this author was honest enough to admit that he didn't know what he was talking about. A reasonable inference, of course, was that some of the objectors to barberry eradication in the United States probably were not much better informed.

This publication also included evidence from much closer to home and from competent people. A quotation from a speech by Professor H. S. Jackson of Purdue University stated that one barberry bush in Indiana was responsible for rust over an area of nearly 1,000 acres, with a total loss of 12,520 bushels of wheat and $12,500; there were "any number" of similar cases in Indiana, in which heavy losses from rust had occurred before barberry bushes had been removed, with no losses from rust after removal.

From this time on, barberry eradication met no serious opposition, and the program continued with an increased budget—$500,000 a year for several years in the mid-1920s. In WPA days in the late 1930s, the budget was one to two million dollars per year. The success of the barberry eradication program is indicated by the fact that in the 1950s some counties in Minnesota each offered a bounty for any wild barberry bush found in the county. They never had to pay; the barberry, wild and cultivated *Berberis vulgaris*, once so common, had been totally exterminated.

The success of the campaign is impressive in view of the difficulty of eradicating barberries. The common barberry is a good competitor. Wherever it has been introduced, in Europe and America, it has quickly escaped from cultivation and established itself in the wild. Its seeds are readily distributed by

The Conference for the Prevention of Grain Rust was formed to co-operate with Federal and State governments in bringing about the control of black stem rust through eradication of the common barberry bush.

Black stem rust originating on the common barberry bush takes an average annual toll of 50,000,000 bushels of grain in normal years, while in epidemic years the losses have been as high as 200,000,000 bushels.

The Conference consists of representatives of the agricultural interests of the thirteen Northern grain growing states, these being the states included in the Federal barberry campaign. The members are the governor, president of the state farm bureau federation or corresponding farmers' organization, state commissioner of agriculture and a representative of the state experimental station of each of the thirteen states, and members at large.

EXECUTIVE COMMITTEE

J. A. O. PREUS, Governor of Minnesota, President
FRANKLIN M. CROSBY, Minneapolis, Vice President
HARRISON FULLER, St. Paul, Secretary-Treasurer
C. W. HUNT, President, Iowa Farm Bureau Federation, Des Moines
J. A. KITCHEN, Commissioner of Agriculture and Labor of North Dakota, Bismarck
C. P. NORGORD, Commissioner of Agriculture of Wisconsin, Madison
J. F. REED, President, Minnesota Farm Bureau Federation, St. Paul

PERMANENT HEADQUARTERS
510 McKnight Building, Minneapolis

This office is prepared to supply in any quantity desired complete information concerning the causes of black stem rust, its effect upon grain and methods of control, including government publications on the subject, posters, specimens, motion pictures, lantern slides, displays for schools, county and state fair exhibits, statistics, etc.

Figure 6. Introduction to the Proceedings of the Second Annual Conference for the Prevention of Grain Rust, held at the University of Minnesota on November 14, 1922.

DELEGATES

IN ATTENDANCE AT THE SECOND ANNUAL CONFERENCE FOR THE PREVENTION OF GRAIN RUST

STATE DELEGATES

COLORADO

L. M. Taylor, Secretary, State Board of Agriculture, Fort Collins.
Dr. A. K. Peitersen, Botanist, Colorado Agricultural College, Fort Collins.

ILLINOIS

Frank I. Mann, representing the Illinois Agricultural Association, Gilman.
O. T. Olsen, Superintendent, Division of Plant Industry, State Department of Agriculture, Springfield.
George H. Duncan, Associate in Crop Production, Agronomy Department University of Illinois, Urbana.

IOWA

C. W. Hunt, President, Iowa Farm Bureau Federation, Des Moines.
I. E. Melhus, Station Plant Pathologist, Agricultural Experiment Station, Ames.

INDIANA

James K. Mason, Director, Indiana Federation of Farmers' Associations, Milton.
Dr. H. S. Jackson, Chief in Botany, Agricultural Experiment Station, Lafayette. (Representing Governor Warren T. McCray.)

MICHIGAN

James Nicol, President, Michigan Farm Bureau Federation, South Haven.
John A. Doelle, Commissioner of Agriculture, Lansing.
W. F. Reddy, State Leader of Barberry Eradication, East Lansing.

MINNESOTA

J. F. Reed, President, Minnesota Farm Bureau Federation.
N. J. Holmberg, Commissioner of Agriculture, St. Paul.
Dean E. M. Freeman, College of Agriculture, St. Paul.
Leonard W. Melander, State Leader of Barberry Eradication, St. Paul.

MONTANA

A. H. Stafford, President, Montana Farm Bureau Federation, Bozeman.

NEBRASKA

H. D. Lute, Secretary, Nebraska Farm Bureau Federation, Lincoln.
Dr. George A. Peltier, Station Plant Pathologist, Agricultural Experiment Station, Lincoln.

NORTH DAKOTA

Governor R. A. Nestos, Bismarck.
Hans Georgesen, President, North Dakota Farm Bureau Federation, Niagara
Joseph A. Kitchen, Commissioner of Agriculture and Labor, Bismarck.
Dean H. L. Bolley, College of Agriculture, Fargo.
George C. Mayoue, State Leader of Barberry Eradication, Fargo.

Figure 7. Conference participants.

OHIO

A. E. Anderson, Director, Grain Marketing Department, Ohio Farm Bureau Federation.

L. J. Taber, Director of Agriculture, Columbus.

SOUTH DAKOTA

W. S. Hill, President, South Dakota Farm Bureau Federation, Mitchell.

Frank M. Byrne, Commissioner of Agriculture, Pierre. (Representing Governor W. H. McMaster.)

M. R. Benedict, Assistant Commissioner of Agriculture and Professor of Farm Economics, College of Agriculture, Brookings.

Dr. Arthur T. Evans, Associate Agronomist and Crop Pathologist, College of Agriculture, Brookings.

Dr. N. E. Hansen, Professor of Horticulture, College of Agriculture, Brookings.

WISCONSIN

George W. McKerrow, President, Wisconsin Farm Bureau Federation, Madison.

C. P. Norgord, Commissioner of Agriculture, Madison.

Dr. S. B. Fracker, State Entomologist, State Department of Agriculture, Madison.

William A. Walker, State Leader of Barberry Eradication, Madison.

WYOMING

Dr. A. F. Vass, Professor of Agronomy, College of Agriculture, Laramie.

DELEGATES AT LARGE

J. R. Howard, President of the American Farm Bureau Federation, Chicago.

Franklin M. Crosby, representing milling companies, Minneapolis.

C. C. Webber, representing implement dealers, Minneapolis.

OTHERS IN ATTENDANCE

Dr. Carleton R. Ball, Cerealist in charge, Office of Cereal Investigations, U. S. Department of Agriculture, Washington.

Dr. F. E. Kempton, Pathologist in charge of barberry eradication, U. S. Department of Agriculture, Washington.

Dr. E. C. Stakman, Plant Pathologist, U. S. Department of Agriculture and Minnesota Agricultural Experiment Station, St. Paul.

Noel F. Thompson, Assistant Pathologist, U. S. Department of Agriculture, Madison, Wis.

Dean W. C. Coffey, Department of Agriculture, University of Minnesota, St. Paul.

Frederic Cranefield, Secretary, Wisconsin Horticultural Society, Madison, Wis.

F. L. French, Secretary, Minnesota Farm Bureau Federation, St. Paul.

R. O. Cromwell, Crop Expert, F. H. Peavy & Co., Minneapolis.

E. C. Leedy, Immigration Agent, Great Northern Railway, St. Paul.

Figure 8. Conference participants.

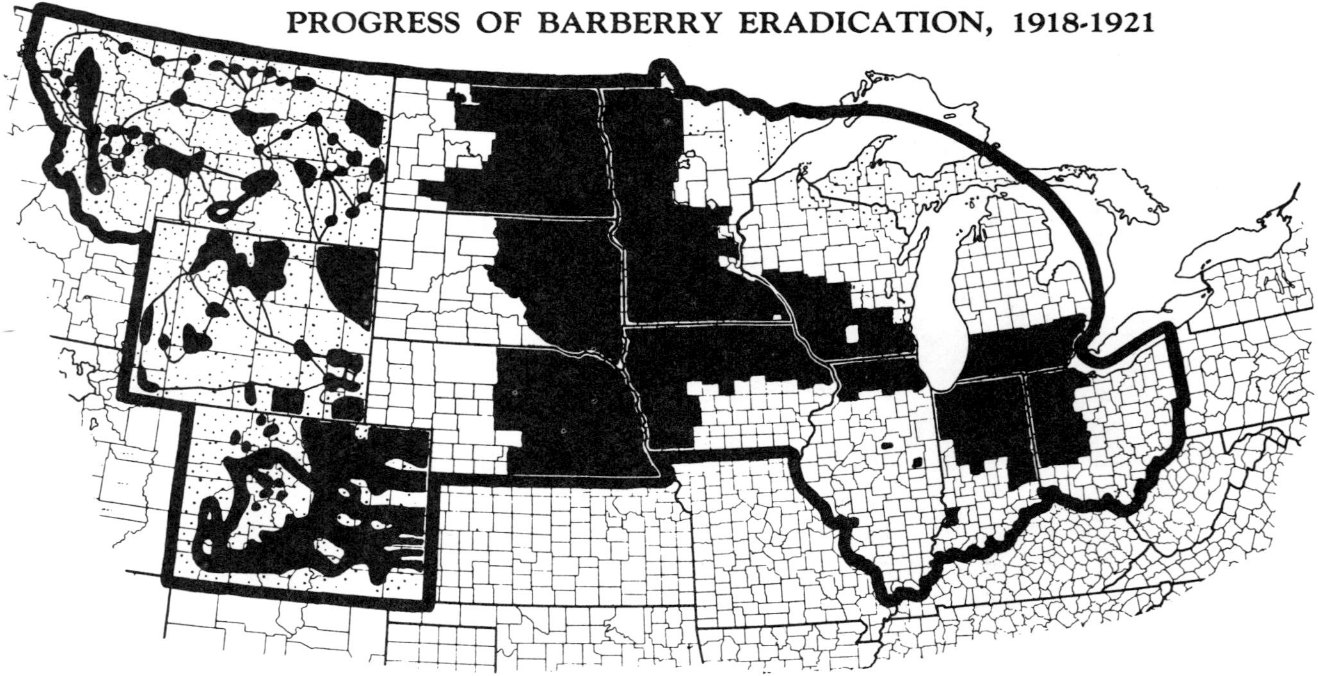

PROGRESS OF BARBERRY ERADICATION, 1918-1921

Figure 9. Map showing the thirteen states in the barberry eradication campaign and the progress made in this campaign by the time of the conference.

60

birds. If a well-established bush is simply dug up, it is not necessarily eliminated, but may be multiplied and propagated, because small portions of the roots left in the ground sprout, and each forms a new bush. Also the soil disturbance may stimulate dormant seeds to germinate, giving rise to many seedlings where one bush grew before. Several thorough goings-over are necessary to make sure that no established bushes have been missed and that new bushes are detected. But the campaign was successful, as Stakman in 1918 was confident that it would be.

Only later, but not too much later, in the early 1930s, did they realize that in eradicating barberries they were doing much more good than they anticipated when the program was begun. The stem rust fungus goes through the sexual stage on the barberry, and it is there that new parasitic races can be produced by hybridization. Race 15B, which was to do so much damage in the 1950s, and which will be taken up in some detail later, was first found near barberry bushes in New York, unquestionably a hybrid. Without eradication of the barberry bushes, stem rust would have been epidemic every year in portions of the 13 states in which barberries were eradicated. As it was, these states went through many years with little rust, then a few epidemic years, followed by 10 years in the 1940s and early 1950s virtually free of rust. These rust-free years were followed by the epidemics of the mid-1950s caused by Race 15B. Those were the last. From 1960 on, no serious epidemics, no heavy losses, from stem rust of wheat have occurred in any of the major wheat-growing regions anywhere in the world. The work on epidemiology of stem rust, begun by Stakman in 1910–1911 and continued so effectively by him and his many co-workers and students, paid off extremely well.

With the eradication of barberries well under way and the continuation of the eradication program assured, Stakman could turn to other problems in the epidemiology of stem rust. One of these was the possible significance, in the development of epidemics, of wind dissemination of urediospores, the repeating spores of the rust fungus. How far could they be carried and still cause infection? From Mexico to Canada? This could be answered by research, and Stakman got the research started.

CHAPTER SEVEN

Aerobiology

In "The Black Stem Rust and the Barberry," in the 1918 USDA yearbook,[1] Stakman wrote,

> The question naturally arises whether the rust spores which overwinter in the South could not be blown northward in the spring and infect the growing grain. In this way the rust might travel from south to north by successive stages as the crop develops. Evidence based on careful observations shows quite clearly that this does not occur. The rust develops on barberry plants in the North and spreads to grains and grasses quite as early in the spring as the rust begins to become general in the South. Then, too, the form or race of wheat rust which is common in the South can not cause rust on hard spring wheats or on most of the hard winter wheats of the North. Even if the rust did blow up from the South, therefore, it could do very little damage in the North.

This was quite true at the time. Once barberries had been eliminated from the northern wheat-growing areas, however, races of rust that *could* infect the winter wheats in the southern Great Plains and the spring wheats in the northen Great Plains and the Prairie Provinces of Canada became widespread and the situation changed. Even before 1915, Freeman, or Freeman and Stakman, began exposing greased slides to catch airborne spores and examining these to detect urediospores of rusts, and the concept of north-south and south-north spread of these spores by the wind, as a factor in the epidemiology of stem rust, was in their minds. The first paper on this was in 1923 by Stakman, Henry, Curran, and Christopher.[2] The introduction states,

> As a part of the rust-epidemiology investigations which have been made since the spring of 1917 by the Office of Cereal Investigations of the United States Department of Agriculture in cooperation with several State experiment stations, attempts have been made to get as much information as possible on the dissemination of spores by air currents and to correlate the data with the spread of rust on host plants. The usual method in studying the distribution of spores of pathogenic fungi by wind has been to expose spore traps of various kinds near the surface of the earth. The possibility of spores being carried to higher altitudes by convection currents, whirlwinds, and other air movements has been recognized, and spore traps have been exposed on high buildings, on mountain tops, and other elevated points. However, it is desirable to know how many spores there are in suspension several thousand feet above the surface of the earth.
> During the spring and summer of 1921 spore traps were exposed on

airplanes in the hope of obtaining more information on the dissemination of aeciospores and urediniospores of *Puccinia graminis* and other rust fungi. The general results seem to be worth recording, although the problem of rust epidemiology was not solved.

Slides were exposed from planes, mostly by means of a gadget that was attached to a wing or wing strut and permitted exposure of six slides, one after another. Each slide was enclosed after exposure to protect it from later contamination, then examined microscopically in the laboratory.

The results of these preliminary experiments indicate that large numbers of spores and pollen grains are carried several thousand feet above the surface of the earth during the growing season. Probably they are carried long distances by the upper air currents, the direction and velocity of which are quite different from those near the surface. If the spores retain their viability, as some of them quite probably do, it is conceivable that a local epidemic might occur in one locality as a result of the blowing in of spores from an infection center in another distant locality.

Slides were exposed on many flights in Texas, Oklahoma, Kansas, and Nebraska. In none of these flights were appreciable numbers of spores caught unless rust was present in fields below where the flights were made. Later they were caught far from rusted fields.

Shortly thereafter, a program was begun in which greased slides were exposed in a holder on a wind vane at selected locations from Texas to North Dakota, mostly by collaborators in country grain elevators. A box of 24 slides was sent to each of these collaborators in the spring. One slide was exposed each day and returned to the box. When the box was full, it was returned to the rust laboratory in St. Paul, and a selected area on each slide was examined for rust spores. It was tedious work, but over the years it established the pattern of wind dissemination of rust spores from south to north through the Great Plains wheat belt. In a 1955 summary,[3] Stakman stated,

Barberry eradication did not eliminate rust epidemics because rust can spread from Mexico, where there are varying numbers of urediospores throughout the year, to Texas, to Oklahoma, to Kansas, to Nebraska, to the Dakotas and on into Canada. During the first week of June 1925, as an example, there was rust on wheat as far north as central Kansas. For almost a week the wind swept northward at 17–26 miles an hour, carrying rust spores 600 miles northward on a front more than 400 miles wide, thus inoculating grain in an area of a quarter of a million square miles. It was calculated that in early June 1952 there were 4000 tons of urediospores (150 billion spores per pound) on 4 million acres of wheat in sixteen counties of northern Oklahoma and south-central Kansas. Winds carried the spores to Minnesota and the Dakotas where they were deposited at the rate of 3.5 million an acre in an area of 40 thousand square miles. On an acre of fairly heavily rusted wheat there are about 50 thousand billion spores. Small wonder that strong winds followed by heavy dews or rain terrify those who have known the red scourge of wheat. In 1954 spores were being deposited at the rate of 10 thousand to 50 thousand per square foot daily at certain places in North Dakota 6 weeks to 2 months before harvest.

Tracing the progress of stem rust from south to north, by means of exposed slides, was continued for many years. A different system, employing small battery-driven whirligigs and cellophane tape, is still used occasionally to sample airborne spores, but the regular surveys to detect and trace the south-north progress of spores have been discontinued. For many years not enough stem rust has occurred on wheat in the Great Plains to make long-distance spread of urediospores by the wind much of a factor in its epidemiology. No general epidemics of stem rust have occurred in the Great Plains in recent times, nor are any likely to occur in the future, thanks in part to the physiologic race surveys that alert plant breeders to potentially damaging new races of rust in time for them to take preventive action.

CHAPTER EIGHT

Physiologic Race Surveys

The hunt for physiologic forms or races of *P. graminis tritici* began, of course, with Stakman's collecting trip to the Pacific Coast in 1916, when he found the new race *P. graminis tritici compacti* that he and Piemeisel later reported on. The hunt continued in 1917, with collections in Oklahoma, Texas, Louisiana, and Alabama. The formal, annual, well-organized surveys, undertaken with the aim of mapping the distribution and prevalence of the major races present throughout the country, began about 1930. The surveys have continued up to the present and presumably will be continued into the foreseeable future because they are an essential part of the never-ending campaign against stem rust.

A long time passed before the significance of physiologic or parasitic races was clear to some agronomists and plant breeders who were trying to develop varieties of wheat resistant to stem rust (and varieties of wheat and other plants resistant to other diseases). Some of them pooh-poohed the different parasitic capabilities of the different races, on various "logical" grounds, although the evidence was there for anyone not blinded by preconceptions to see. Some claimed that the different parasitic races were only a greenhouse or laboratory phenomenon, although the races obviously had been collected in the field, where they were causing rust, often severe rust, on varieties that had been selected as resistant to rust. Some of this opposition was vociferous, probably on the principle that the less solid your stand, the more loudly you must maintain it, and some of the opposition lasted up into the mid-1930s and beyond. Some of the rust-resistant varieties of wheat released up until the early 1940s were widely proclaimed as permanently resistant, although they were later severely damaged by Race 15B. So far as can be seen now, this widespread and totally erroneous belief in the unimportance of parasitic races, and the opposition to accepting it as an essential ingredient in the plant breeding procedure, was simply a reluctance on the part of those concerned to learn something new, especially from a plant pathologist, and a fairly young one at that. If Stakman was sometimes rather belligerently right about this, maybe he had to be; it was an important principle, which had to be understood as soon as possible for the good of agriculture. He told some of the history of this in a seminar given in Sonora, Mexico, on April 25, 1977, three weeks short of his 92nd birthday.[1] The terrible stem rust epidemic of 1916 had eliminated the variety Marquis from the spring wheat area of the Upper Midwest, and a replacement was being sought.

> The other consequence was the elimination of the variety Marquis, creating the question: What could serve as a substitute? There was nothing to substitute, and so we became a durum-growing area. Temporarily, at least, rust had eliminated the bread wheats from the area [the Upper

Midwest]. From 1917 to 1923 you could not find any fields of bread wheat in the entire area. All durum, nothing but durum. Now, switching to the durums brought its own troubles, too. The durums were much more susceptible to certain diseases, particularly ergot and root rots, than the bread wheats had been. So we were just changing diseases again.

But the durums lasted only until 1923. In 1923 heavy rust appeared and when we isolated from it we found a preponderance of Race 11. Actually we had [identified] about 22 races by that time, even though the 11–32 complex appeared in 1923. And then the question was, what was there to substitute for the durums? There had been three lines of progress up to that time. One of them was the development of the variety Marquillo. This was a cross between Marquis and Iumillo durum. Now, that was thought to be a big step forward because there was supposed to be a linkage between durum characters and rust resistance. It was freely stated that the linkage could not be broken. But out of some 2,200 lines there were three which had bread wheat characters and the durum resistance. Thus by using large numbers the linkage had been broken.

In the meantime a hard red winter wheat called Kanred had been selected in Kansas. It was described as a high quality stem-rust-immune variety. But in the very year when that was described, Dr. Drexler and I were down in Oklahoma in the fall to see how much rust there was. We found very large pustules on Kanred, and wheat in general. When we isolated from many of these pustules, we found Race 3. In other words, Kanred was immune to some races but completely susceptible to others that were already in the area.

In the meantime, a cross had been made between Marquis and Kanred to transfer the spring wheat character into a variety that was resistant, called Reliance. A cross of (Marquis × Iumillo) × (Marquis × Kanred) produced Thatcher.

There was still another source of resistance at this time and that was Kota, which was a resistant bread wheat selected out of a heavily rust[ed] field of durum. But when they tried to grow Kota it was terribly susceptible to leaf rust, loose smut, stinking smut, and to many other diseases.

But hope springs eternal. In North Dakota they crossed Kota with Marquis because the latter variety had such good milling qualities, and obtained the variety Ceres. [This was in 1926.] Ceres was a pretty good wheat although it had a lot of susceptibility to diseases. However, when grown in the area in which it was grown it did produce fairly well. I remember I suggested once at a meeting that they should not rely on Ceres for very long because it was susceptible to certain races of stem rust. I was almost thrown out of the room. I thought they were going to banish me to some desert island. I managed to rub it in later and say: "I told you so."

Ceres was distributed in 1926. In 1928 we found race 56, to which Ceres was very susceptible. The race increased slowly for a few years, and then all of a sudden exploded. In 1935 Ceres was ruined along with all ordinary wheats. In the minds of a lot of people, including its promoters and seed sellers, it had been a perfect wheat.

The first collection of the then new race 56 was made in Dr. [Norman] Borlaug's native state of Iowa, a little west of Des Moines. It came from barberries, so without any question the evidence was that we now had a hybrid between certain rust races. And so we began to find that barberry was indeed not only a source of rust but a source of new races of rust through sexual recombination.

But in the meantime there was another source of resistance named Yaroslav Emmer. Yaroslav Emmer crossed with Marquis gave rise to the varieties Hope and H44. Breeders at that time offered a dollar for every pustule of rust that anybody could find on these two new varieties. This proved to be a very poor offer, since they soon found that these varieties were very susceptible to certain other races.

There was still another line of progress made, using Thatcher... crossed with Hope. The F_1 was backcrossed twice to Thatcher. A selection from the second backcross, named Newthatch, was increased and distributed.... From Newthatch and from Yaroslav Emmer... we obtained a certain degree of insurance against the destructive rust disease (i.e., generalized resistance and slow rusting, especially in adult plants). Between 1935, when these materials were becoming available, and 1950, with the eradication of many barberries, we had the golden age of wheat production in the U.S.A. There were only four races prevalent, and ninety percent or more of the inoculum at that time was of those four races.

Even during these years of scarce stem rust, Stakman was writing and speaking and publicizing the idea that stem rust might strike again, but by 1950 some of the perennial optimists again were proclaiming that the days of severe stem rust epidemics were over, a thing of the past. Stem rust had been conquered. Then Race 15B got loose.

Race 15B had been found occasionally in the rust surveys from 1940 on, mostly near barberries. It was known to be potentially dangerous because all of the widely grown bread and durum wheats in the United States were susceptible to it. If it became widely distributed, and weather favorable to rust development prevailed, it could become destructive. Because of a peculiar combination of winds and weather in 1950–1951, Race 15B did become widely distributed throughout the wheat belt from northern Mexico through the Great Plains of the United States and on into the Prairie Provinces of Canada. Weather favorable for rust development came in 1953 and 1954, and there were devastating epidemics—in millions of acres in the Upper Midwest, the wheat crop was virtually a total loss.

Work on development of varieties resistant to Race 15B had been under way for some time by then, and these new varieties soon replaced the susceptible earlier ones. Since that time there has been no widespread or seriously damaging epidemic of stem rust on wheat in the United States, Canada, or Mexico (or any of the countries where the wheats produced by Borlaug and his team [from the International Center of Corn and Wheat Improvement] are grown). The worldwide, 60-year war waged by Stakman and his collaborators against stem rust has been successful to the extent that this ancient scourge has been reduced from a major to a relatively minor menace. As in all wars, there were a lot of foot soldiers who did the dirty fighting and who received little recognition and won no medals, but Stakman was a pretty good general, too. He not only planned and directed the overall strategy, but also spent a good deal of time in the laboratory (Figure 10) and out in the field where the action was.

Variation, Mutation, and Hybridization in the Smut Fungi

Due at least in part to the work by Stakman and his colleagues, collaborators, associates, and students on parasitic races of the stem rust fungus, the idea that fungi and other microorganisms might be variable, and not static, was becoming recognized by 1920. Some genetically minded mycologists and some geneticists with mycological bents were beginning to find out that fungi, like morphologically more complex living things, had heritable characters. This should not have been very surprising, since by then all of the fungi that had been investigated had been found to have nuclei more or less like those of higher plants and animals, with chromosomes, and at least some of them enjoyed the biological benefits of sex. A German, Hans Kniep, was publishing research papers by 1919 on sexuality in fungi, and in 1928 he published a book called *Sexuality of the Lower Plants* (*Die Sexualität der niederen Pflanzen*)[2] that was required reading for students in Stakman's department. Stakman, of course, had helped construct this bandwagon of variation and variability in fungi, and he climbed aboard and helped drive it with characteristic gusto. He chose to work mostly with plant-parasitic fungi rather than with economically unimportant saprophytes. The principles could be established with either, but it was characteristic of him to choose as his research subjects fungi important to the region where he worked and important to agriculture in general. Among these were several species of smut fungi that cause diseases of cereal crops. One of these smut fungi, *Ustilago maydis* (or *U. zeae*), causes smut on corn and is somewhat unusual in that it can be grown readily on agar media, where it forms colonies with characteristic topography and color (usually, only the 1n or haploid phase could be so grown). The fungus also could be hybridized, and then the haploid progeny could be studied in culture and the diploid progeny in the corn plant. Work with mutation and hybridization in *U. zeae* began about 1924 and

Figure 10. Stakman in the rust lab in the old Tottering Tower.

continued into the 1940s. This work with corn smut and other smuts served as material for Ph.D. theses for dozens of graduate students from many countries. As Stakman tells it in the oral history:

> I'd like to make it clear that we selected the smuts for concentrated study for two reasons, first because of their economic importance, and second,... because they have certain characters that made it possible to study [them] most easily and get the clearest results on the things that we wanted to know about in connection with the production of new biotypes, new physiologic races. That's why we selected them.

And as to corn smut:

> I think that corn smut may be somewhat more variable than certain other organisms. We know that it is. And we also know this, that there is a difference in the mutability of different biotypes or lines within the corn smut organism, because we made mutable times mutable crosses, constant times constant crosses, and we fixed the constancy by inbreeding over a number of generations; after the fifth or sixth generation we got constancy.[3] After the sixth generation approximately of crossing mutable by mutable lines, we got something that was so mutable that we couldn't keep track of the original line in the mixtures. Now, when you cross[ed] mutable lines by constant lines, you got segregation for mutability and for constancy, and so the conclusion naturally is that there are genes for mutability and there are genes for constancy. The degree of mutability within the species itself is determined by the genes in these individual lines.

Concerning some of the other cereal smuts, he felt that

> the most significant thing perhaps about some of the other cereal smuts like those of oats and of barley and of sorghums and of millet and that sort of thing was that there it was easy to demonstrate not only intraspecific hybridization, but also interspecific hybridization. There are a lot of people that used to define a species as something that wouldn't cross with something else, but actually different species which are good morphological species can really cross with each other. I think that's generally conceded now, but that used to be in the definition.

One of the benchmark papers on this subject of variation in the smut fungi was by Stakman and his associates in 1930.[4]

Some investigators continued to quibble about whether what Stakman and his co-workers on the smut fungi called mutation actually was mutation. This depended in part on how one defined the term, but if, as Stakman said, mutation is defined as "an inheritable change which arises by means other than hybridization," then what they were calling mutations were mutations. As always, there were some to whom the definitions were more important than the phenomena. In any case, over a period of almost 30 years, with most of the students and staff in plant pathology at Minnesota devoting most of their research to variation in phytopathogenic fungi, the principle that fungi are varying and variable became well established, and Stakman played it up on the lecture circuit under the catchy title "Plant Pathogens Are Shifty Enemies."

PART III

Educator

Stakman in his office in 1949. (Copyright 1949 by the American Association for the Advancement of Science. Reprinted, with permission, from "The Ninety-Eighth President of the AAAS," J. J. Christensen, *Scientific Monthly* 68:31-34, January 1949)

CHAPTER NINE

Teacher

Scholarship and teaching ability do not always go together, but in Stakman they did; he was unusually gifted in both. He combined wide intellectual and scientific interests with an unusual capacity for learning and with an extremely retentive and accurate memory. Along with this was thoughtfulness, the ability to analyze facts and combine them into meaningful concepts and to synthesize ideas and principles. He wasted very little time on recreational or junk reading, and his mind was always in gear and working, never idling. In teaching he had the very rare ability to make almost any subject he touched seem interesting, alive, and important to his listeners, and he made students feel that they were fellow adventurers exploring with him newly discovered worlds where truth and reason and order prevailed. He could even give an analysis of the subjunctive mood in grammar that made it appear simple, lucid, straightforward, and interesting.

He continued to learn and to teach up until his final illness. To him, learning and teaching were as normal, natural, and necessary as breathing. Most of his teaching, after he received the Ph.D. degree in 1913, was at the University of Minnesota, but he taught for a semester at the University of Halle-Wittenberg, in Halle-an-der-Saale, Germany, and later, at various times, at the National School of Agriculture in Chapingo, Mexico, and at the Agricultural School at Saltillo, Mexico. In Germany he lectured in German and, typically, found it difficult to stop when the time was up; in Mexico, he lectured in Spanish. He talked and gave seminars and shared his thoughts on various subjects in many places in many countries; an audience of two or more and a time span of at least 15 minutes sufficed for Stakman to give a moderately profound analysis of a current or historical problem or situation, and sometimes an audience of less than two and a time span shorter than 15 minutes sufficed. This chapter is an attempt to summarize some of his more formal teaching over the years.

Undergraduate Courses

Esther Tolaas's recollection of Stakman's teaching of household bacteriology to women in home economics in 1913 has already been quoted in Chapter 1. Even after more than 60 years, she still vividly remembered him as "an attractive young man" and as "a hard taskmaster." Perhaps he was a demanding teacher, although almost any teacher who has anything fairly solid to offer and is moderately insistent upon having the students learn a reasonable portion of it is likely to be considered a hard taskmaster by some of them. Louise Jensen, who was to become Mrs. Stakman, taught one of the laboratory sections in the household bacteriology course, and Arne Tolaas taught the other. The home economics students in the course were sent to the Dairy Department to learn bacteriology in relation to milk and also spent time in the Veterinary Science Department, which probably was a Stakman innovation and which must have opened the eyes of some of the students to the practical aspects of bacteriology; the course was more than just book learning. Stakman probably learned much about bacteria and bacteriology too; a good teacher usually learns more about the subject being taught than the students do.

In the fall quarter of the 1914–1915 academic year, Stakman taught a course in wood structure and identification to students in the School of Forestry of the College of Agriculture, Forestry and Home Economics. He had studied plant anatomy in his botany undergraduate major and so had enough background to keep ahead of the students. Some of his lecture notes for this course still exist, of which a sample in his own handwriting is shown in Figure 11. The course as he taught it probably had a flavor of general plant anatomy, with emphasis on the structure of woody plants and forest trees, and although forestry students then and for quite some time after objected, sometimes vociferously, to anything that they considered not "practical," Stakman evidently did not hesitate to have them learn about cells and tissues in squash plants and grasses. He probably even made them feel that it was important to do so. Judging from the exams given at intervals during the quarter, a representative sample of which is shown in Figure 12, it was a course of substance and the students were expected to learn plant anatomy. In the final exam, they also had to identify about 100 samples of different kinds of wood, so there must have been a laboratory portion of the course. Whether that was taught by Stakman or by someone else is not known; most likely it was by someone else, with Stakman overseeing it.

Stakman was a gifted teacher, but his preference was for the expository approach, with an assistant to handle the time-consuming and back-breaking drudgery of the laboratory. He recognized the value of laboratory work in teaching natural science, and when some of his staff members later developed outstanding "hands-on" laboratory and greenhouse work for the undergraduate courses in plant pathology and forest pathology, they had Stakman's full approval. However, he did not back up this approval by providing much in the way of financial assistance for microscopes or other equipment and supplies. Until after Stakman's retirement, one stereoscopic dissecting microscope, of about 1935 vintage, was available in the department for all undergraduate and graduate teaching. The ordinary high- and low-power classroom microscopes were in such short supply that frequently the instructor in charge of the undergraduate course in plant pathology or forest pathology had to arrange on his own limited authority the loan of microscopes from some other department on the campus. If the instructor in forest pathology wanted specimens of stem cankers or trunk or root rots of forest trees, he could go up to the North Woods a couple hundred miles away in his vacation time and get them, but not at departmental expense; he paid his own way. That is, Stakman approved of laboratory work, but he did not personally take part in it or put his money on it.

He loved argument for its own sake, and he sometimes carried it to extremes, as did medieval scholasticism, in which the question of how many teeth a horse has was settled not by examining the mouths of half a dozen or half a hundred horses, but by discussion and logic. As an example, in a seminar dealing with fungus stains of wood, Stakman and one of the older graduate students, a forest pathologist by profession, argued for some time and with some vehemence as to

①

Wood Technology. Sec. 4   10/1/14

Secondary Thickening of Stems.

   I Review.
      Primary meristems
      Development of primary permanent tissue systems.

   II Development of Secondary tissues.
      A. Formation of Cambium
        a. fasicular
        b. inter fasicular.

      B. Activity of Cambium
        a. Xylem – more xylem cells than phloem.
        b. phloem
        c. New medullary rays. – Secondary.

      C. Phellogen o. Cork Cambium
        a. Origin
          1 Epiderm
          2 Cortical parenchyma
          3. pericycle.
        b Activity
          1. Cork – note lenticels.
          2. Secondary cortex.

      D. Annual Rings.
        a. Sequence of rings
        b. Cause of appearance.
          1. Spring – need for new Conduction
          2. Fall – " " " Support.

      E. Heart & Sap
        Duramen & alburnum.

Figure 11. Notes for Wood Technology, in Stakman's handwriting.

Wood Tech.   Quiz 11/17/14.

I.

a. Describe ray tracheids; discuss their origin & distribution
b. Distinguish between the following on the basis of ray tracheids:
    1. Pinus Strobus + P. resinosa
    2. " " + P. palustris
    3. P. palustris + P. resinosa
    4. Thuya occidentalis + T. japonica.
    5. Thuya + Tsuga.
    6. Tsuga + Larix ✓
    7. Larix + Taxus
    8. Juniperus + Sequoia
    9. Taxodium + Thuya
    10. Picea + Abies.

II

Discuss resin cells in the following genera:
    1. Abies
    2. Tsuga
    3. Thuya
    4. Taxodium
    5. Larix
    6. Juniperus
    7. Sequoia
    8. Libocedrus
    9. Pinus
    10. Pseudotsuga.

III

a. Classify medullary rays on the following bases; Describe each in detail
    1. Composition
    2. Structure

b. How can the following be distinguished on the basis of med. rays?
    1. Salix + Populus
    2. Fraxinus + Ulmus
    3. Ulmus + Celtis.

IV.

Contrast hard + soft pines in every way possible

Figure 12. A quiz for Wood Technology.

76

whether the green stain, caused by the fungus *Chlorosplentium*, in a particular sample of wood was superficial or extended deep into the wood. When an impatient class member finally split the piece of wood to show that the stain extended throughout the interior, this was brushed aside as a somewhat spoilsport and counterproductive ending of what had been a stimulating intellectual exercise. This may have been an extreme case, but not too extreme; with Stakman, the intellectual exercise sometimes took precedence over demonstrable fact.

Graduate Courses

The courses that were taught regularly by Stakman and that all or nearly all graduate students in plant pathology took at some time during their graduate career were History of Plant Pathology, Principles of Plant Pathology, and, from the early or mid-1930s on, Genetics of Plant Pathogens.

The course in history consisted of weekly lectures. In the 1930s, they were in the evening, each session running from 8:00 to 11:00 or beyond, which may have strained the attention span of the students, some of whom had little predilection for history. Later, in the 1940s, they were from 4:00 to 6:00 p.m.

The lectures were given on the fourth floor of the old Tottering Tower (Agricultural Botany Building) in the long, narrow, and moderately intimate seminar room, where hard-bottomed student chairs with arm rests were arranged along three of the walls. A long, narrow table ran up the middle of the room. Several small half-windows gave out onto a flat roof. It was not an airy, spacious room but rather cramped and confining. Smoking was permitted, in fact encouraged, Stakman of course smoking one of his 200 or 250 pipes, with relighting being required every minute or two. Some of the students also puffed away diligently on pipes (although few continued pipe-smoking after they received their degrees and left); some smoked cigars—or even cigarettes, which Stakman considered effeminate. The air soon was thick with smoke, and, because of the very limited ventilation, it remained so and became thicker as the evening wore on. To some of the nonsmokers it must have been poisonous.

For some of the history lectures, there were mimeographed lists of references. The course may have developed in part from a seminar series in 1917, in which the students worked up papers on outstanding botanists, biologists, and plant pathologists of the past, but, from his undergraduate days on, Stakman had been interested in history—ancient, medieval, and modern; political, social, agricultural, and scientific. He knew history, and he could make it come alive. The emphasis throughout was on the development and evolution of ideas and concepts and on the scientists and their times, and Stakman could present it with a wealth of human interest and detail, as if he personally had worked with these people and had talked and thought with them. It was, for anyone with a liking for history, a wonderfully stimulating and informative course, and it formed a solid base for those who chose to go on, on their own.

No exams were given in this class—nor were they in most of the other classes that Stakman taught. Everyone received a grade of "S," for "satisfactory," a grading system that Stakman continued to follow long after it had been officially banned by the Graduate School. (They quietly changed the "S" grades to "B" on the students' records, which irritated some of the students who considered any grade less than "A" on their graduate school transcripts to be equivalent to a felony conviction on their vitas.) Stakman always believed that he was training adults to think, that his, and his department's, major obligation was to turn out professionals of as high a quality as the incoming material permitted, and that the quality of a person who was awarded a Ph.D. degree could not and should not be measured by grades and grade point averages and course credits. He deplored the growing emphasis in the Graduate School on such measurements, considering it to be misplaced. Stakman was training, or trying to train, adults, and he thought that his judgment of their progress, after he had associated and worked closely with them for several years, was better than that of Graduate School clerks or Graduate School computers, or even of Graduate School administrators. He himself had become a plant pathologist without ever having taken a course in plant pathology. Probably none of his students ever equaled that, but some came close; Stakman was no stickler for Graduate School regulations.

Stakman taught Genetics of Plant Pathogens in alternate years. The 1948 class had 37 students, 18 from the Department of Plant Pathology, 14 from the Department of Agronomy and Plant Genetics, four from the Department of Entomology, and one unaffiliated. All of these received the grade of "S." Up until some time after Stakman's retirement, most plant pathology majors minored in agronomy and plant genetics, and most of those who majored in agronomy and plant genetics minored in plant pathology. There was always a sprinkling of students with minors in entomology and in forestry. The plant pathology-agronomy and plant genetics partnership stemmed from the early days of breeding improved varieties of crop plants; one of the important characteristics sought was resistance to plant diseases, and one way to achieve this was to train people in the two basic disciplines.

Seminars

The dictionary defines seminar as "a select group of advanced students associated for advanced study and original research under the guidance of a professor." Stakman did not invent the seminar, but he used it very effectively as an important part of the teaching and learning process. There were two formal seminars in the Department of Plant Pathology (and innumerable informal ones, as discussions were encouraged as a way of life). One was on Tuesday afternoon from 4:00 to 6:00. Graduate students registered for this each quarter throughout their graduate careers and received one credit per quarter. Each year each registered student (nonregistered major and minor students were welcome) presented a paper that he had worked up, usually and preferentially but not necessarily related to his own field of research. Whether or not the subject dealt with his own research, the student was encouraged to contribute some original laboratory research to the paper. The student prepared a text, with references, which was gone over with some care by a seminar committee of elders and, if necessary, revised in accordance with their recommendations; the preparation of one of these seminar papers was not at all a casual business. Some of the papers ran to three or four pages of single-spaced typing, with a bibliography of up to 50 references, all properly cited, following the style of the professional journal *Phytopathology*. Not uncommonly the references included papers in French or German. A well-prepared seminar paper required many hours of reading, work, and study, and many of these papers constituted thorough professional or semiprofessional summaries of the subjects they contained.

The student presented the paper orally to the seminar; reading it was absolutely forbidden. Anyone could interrupt with a question or comment at any time, and the speaker was expected to be at least moderately critical of any facts, opinions, and ideas in the paper. False doctrine was open to questioning and refutation, which of course is the whole idea of a good seminar. A careless, slothful, or indifferent student who did not bother to check up on the material being presented was in danger of ending up covered with confusion and disgrace in front of his peers. This encouraged serious study and reduced bluffing and blow to a minimum. Stakman presided, smoking one of his pipes (Figure 13), from a chair near the speaker and contributed to the discussion both during and after the presentation, especially after, when just about anything discussable could be taken up. No feet were shuffled on the wooden floor toward six o'clock (the students in German universities do this to indicate disagreement with something the professor has just said; it is very unnerving)—anyone whose intellectual hunger was not strong enough to banish all thoughts of dinner was an

Figure 13. Stakman collected pipes, and one was never far away when he talked.

intellectual weakling and a lightweight. The seminar often ran well past 6:00. Stakman once admitted, or maybe boasted, that he had absolutely no sense of time when he was talking, a characteristic not always shared by his listeners. By 6:00 or 6:30, many of the scholars had absorbed about all the wisdom they were capable of absorbing that afternoon and were impatient for their starches and proteins, and it probably would have been better for all concerned just to stop. Stakman never ended a seminar or discussion with anything but reluctance. In any case, the students in this seminar got practice in reviewing literature in a subject, in writing up a paper with a bibliography in acceptable and approved form, in presenting it effectively, and in discussing it with their peers. It was excellent training. And no seminar over which Stakman presided or in which he took part was ever perfunctory or boring.

Each student usually prepared one paper per year, and after attending these seminars every week for three or four years, with all the attendant discussions and questions and ideas, he or she had absorbed a good deal of knowledge of plant pathology and of some related subjects.

The other seminar, which began at 8:00 on Thursday nights, was a sort of informal literature review—mostly current literature in the biological sciences. For a while in the 1930s, each member of the seminar was assigned specific journals to look over and report on, in case anything of special interest was found worthy of report. The chairman was appointed by Stakman and usually served for a year, although I think that C. S. Holton served as chairman throughout most of his graduate career, probably because he was so good at it. Theoretically, attendance at seminars by staff and graduate majors was optional, and in the oral history Stakman so states. However, if anyone missed a seminar, he or she would hear of it the next day; for the staff, attendance at seminars was a condition of employment. Seminar participants sat in the same chairs lining the same walls of the same room in which the Tuesday afternoon seminars were held and were called on one after another for any contribution they might make about literature they had reviewed. Some of the contributions, naturally, were poorly organized and poorly presented, rambling, diffuse, noninformative, and of absolutely no interest or significance; some were perceptive, analytical, concise, and highly informative; others were in between. The good student could learn from all of them. About 9:30, coffee and light refreshments were served, after which the seminar resumed and continued until closing time at 11:00. The discussions were informal and often spirited. Stakman joined in most of them sooner or later, usually sooner. There were never any standoffs or draws in the arguments between Stakman and any of the members because Stakman was umpire as well as player, and so he had the final word on all discussable matters and won all arguments. He had remarkable staying power and could wear down even the most durable opponent.

At times, when the discussion following the literature summaries began to flag, as it sometimes did, Stakman would gradually lead by what seemed to be perfectly reasonable steps to some other topic totally unrelated to anything brought up that evening and would expound on it with an impressive display of erudition and insight. Sometimes these may have been extemporaneous spin-offs of his remarkably wide learning, but sometimes they were practice runs for speeches he was to give. If he did not always hold the audience spellbound with these outflows, at least he held them until 11:00. There is no record of a Stakman seminar, anywhere, having ended early; the seminar members might sometimes be exhausted, but the topic never. Seminars for hundreds of years have constituted an important and essential part of the educational process. Probably few seminars, anywhere, have had the vitality, vigor, and the intellectual nourishment, excitement, and adventure of those presided over and led by Stakman. When old-timers gather to reminisce and lie about their days in the Tottering Tower, they still vividly recall the Stakman seminars, which helped convert innocent and ingenuous boys and girls into at least sometimes thinking and thoughtful adults. As a leader of seminars, formal and informal, Stakman was without equal.

After Stakman's death in 1979, the Department of Plant Pathology devoted an issue of its publication *Aurora Sporealis* to reminiscences about him, written by people who had known him in various capacities. The following comments on his teaching were from former students.

Stakman was pre-eminent as a teacher. In my experience he never bored a class and the skill with which he jollied a seminar along in what we came to regard as his "warm up" exercises was really something to witness. When the main business came everyone was relaxed, receptive and willing to contribute. The dumb did not remain so when he was around. He was so interesting because, like his contemporaries, Buller, Whetzel and William Brown of Imperial College, London, he was a very widely read man, well acquainted with the literature and cultural heritage of his science. Because of this he could make at least some of his listeners almost feel the physical presence of such stalwarts as Oscar Brefeld and his smuts growing in his beerwort and the scholarly Tulasne brothers producing their beautiful drawings of fungi. I found his History of Plant Pathology lectures a delight and I have never parted with my notes on them. As I read I can see him before me performing just as it was. [by J. H. Western][1]

Stakman was a powerful teacher; perhaps at times even overpowering given his massive body of knowledge, his motivation and zeal as a teacher and, beyond all else, his charisma. He was, in a sense, a demanding teacher. He demanded that we think about things and think them in a logical way to a logical endpoint. He insisted that we be curious. He was quick to remind us that scientists do not create data; they merely are observers of biological phenomena. The significance of what we observed would be dictated by our ability to interpret what we observed. It was not a question of what it was, but rather what it meant. [by R. R. Nelson][2]

In September 1930, Stakman, accompanied by Mrs. Stakman, went to Germany to lecture in plant pathology for a semester, in German, at the University of Halle. This arrangement was made by him and Dr. Theodor Roemer, who was head of the Department of Plant Pathology and Plant Protection at the University of Halle. Stakman arrived in Halle only a few days before the first lecture was to be given. In the oral history, he says,

...and so I started to write the lectures out, because I had very little experience in talking scientific German recently, and a two-hour lecture in German seemed like a big undertaking. So I wrote out quite a little of the first lecture. And they had the lecture desk all banked with flowers, and every blooming professor and student in town almost was there for that first lecture, because there was a foreign professor and this was their way of trying to make a foreign professor feel at home, and so they were welcoming me. But at any rate, I started reading, and I hadn't been accustomed to reading lectures. Over there, they read their lectures mostly and they call lecturing "Lesen," reading, and if you teach at a university you "read" there. Well, I got to the point where it seemed to me that when I was talking about "Krankheitserscheinungen" and "schmarotzende Pilze"... my vocal organs didn't function too well, and so I put the paper aside and just talked. . . . But at the end of the lecture,[3] they told me that they had wondered how I was going to extricate myself from some of those complex German sentences. But they said that I had successfully extricated myself from all of them and that within a few days I'd be talking German like a German. Well, it is true that in a very short time it was very natural. Obviously, I worked at it. I tried to talk no English whatsoever, much to Mrs. Stakman's disgust, and I soon got to the point where I felt quite at home in German.

The laboratory portion of the course, with the legwork and backwork done by his laboratory assistant, was also a great success—it was the first time the students had ever come in hands-on contact with diseased plants and living fungi from the surrounding fields and farms and markets.

Stakman also lectured to students in agricultural schools in Mexico. Dra. Ma. de Lourdes de Bauer, a Mexican plant pathologist who received her Ph.D. from the University of Minnesota under Stakman, recalled in the Stakman Day issue[4] of *Aurora Sporealis* that

his association with educational institutions was most profitable to Mexico. His advice was of a great help to the "Colegio de Postgraduados" in Chapingo, Mexico when it established the master of science programs in 1959, as well as in 1973, when the Ph.D. programs became a reality.

Aware of the difficulties in starting the plant pathology graduate program in Chapingo, Dr. Stakman taught his famous course "Principles of Plant Pathology" for several years. At the same time he collaborated with Dr. Alfredo Campos in different ways to get the Department of Plant Pathology well settled. Not only Chapingo was favored by Dr. Stakman. He was also a visiting professor at other educational institutions, such as those in Monterrey, N. L. and Saltillo, Coah.

The Stakman Day issue also included quotations from Stakman,[5] among them one on educational sins.

There are many other educational sins, related both to concepts and methods: the sin of requiring knowledge without understanding; the sin of encouraging absorptive rather than productive scholarship; the sin of killing curiosity; the sin of professorial volubility and student silence; the sin of making verbalism epistatic to sense realism; the sin of the professors who want only research grants but do not want to teach; the sin of the universities that emasculated their colleges of agriculture; and others too numerous to mention.

The phrase "epistatic to sense realism" doesn't make much sense; it doesn't sound like Stakman. To the sin of professorial volubility in the above, Stakman would have had to plead guilty, at least occasionally. Of course in most of the teacher-student situations, Stakman was the teacher, and the students were the listeners, and Stakman probably rightly figured that the listeners couldn't be learning much when they were talking, or that they were more likely to be learning something when he was talking than when they were. A common complaint was that in small social gatherings Stakman often dominated the conversation so completely that no one else could get in more than an occasional word. True. But even his casual talk was likely to be more informative and more entertaining than that of others.

Stakman's Staff in Plant Pathology

Dr. Henry D. Barker, in the Stakman Day *Aurora,* says, "In early 1919, the graduate school majors in Plant Pathology consisted of Julian Leach, Louise Dosdall, and myself." Both Leach and Dosdall became full-time staff members. Leach left in 1938 to become head of the Department of Plant Pathology, Bacteriology, and Entomology at West Virginia University in Morgantown; Miss Dosdall stayed on until she retired in 1956. The next full-time staff members were J. J. Christensen, who received his Ph.D. degree under Stakman in 1925, and Helen Hart, who received the Ph.D. degree under Stakman in 1929. M. B. Moore (M.S. 1934), C. J. Eide (Ph.D. 1934), and C. M. Christensen (Ph.D. 1937) were half-time or full-time staff members before they received their degrees, and they stayed on as full-time staff members until their retirements. Later T. Kommedahl, M. F. Kernkamp, and T. H. King, from the Midwest, E. G. Sharvelle, from England, and I. W. Tervet, from Scotland, joined the staff as full-time members. Sharvelle and Tervet subsequently left. All of these men had received most or all of their graduate training, and all had received their Ph.D. degrees, in Stakman's department. Most or all of them had some research experience elsewhere, but all were pretty well "Stakmanized."

CHAPTER TEN

Department Head

There was no doubt or question in his or in anyone else's mind that Stakman was Head of the Department and that it was his department, which was normal for the times. In the early and mid 1930s, the depression years, when university salaries and staffs were being reduced and hard times were stalking the land, the president of the university, Lotus Delta Coffman, once stated that as long as the university could retain its outstanding department heads, what happened to lesser ranks was of little importance or consequence. This did nothing to cheer the lower ranking staff members, but it was the way things were at the time. So far as his department was concerned, Stakman was the Big Chief (and was often so called), and both he and his staff members knew it. In those days most department heads were in total charge of their departments. Nobody fretted about it or felt held down; it was accepted as the normal and right and reasonable arrangement. One result of this was a total absence of cliques or cabals or attempts to set up separate fiefdoms. Each staff member discharged his or her responsibilities in teaching, research, and extension; attended seminars; and served on various committees that handled departmental or campus affairs. They were not asked for much input on day-to-day running of the department or on budget matters, nor did they feel that administering the department or the university was part of their function. In their own teaching and research, they had just about complete freedom.

There was, in addition, a special ingredient in the Department of Plant Pathology that reduced all other considerations to relative insignificance, and that came from Stakman himself. It was the atmosphere of excitement, accomplishment, and intellectual adventure that prevailed. Stakman had a way of making his staff and students feel that they were on the cutting edge of scientific progress, that their work was important, that their contributions were significant, and that they were on an outstanding team. By the early 1930s, Stakman had traveled extensively in Europe and knew many of the outstanding plant pathologists and botanists in the universities of most of the countries there. He also had been twice around the world and had visited and discussed problems with men in Australia, India, and Egypt. He also had traveled extensively in the United States and Canada and fairly extensively in Mexico, and wherever he had gone he had become acquainted with the leaders in plant pathology and agriculture in general. The work on stem rust done in his department and under his direction was known worldwide. Graduate students from all over the world were coming to Minnesota to study under him; at any one time the department might have graduate students from many states and from Canada, England, Germany, Hungary, Australia, New Zealand, China, and India. From about 1930 to 1941, there were always several graduate students from China and India in residence. It was a cosmopolitan department, and just about everyone was working on one or another aspect of variation in phytopathogenic fungi. They were contributing to the development of a new field, one with important basic principles and a multitude of practical applications in world agriculture. Thanks in good part to Stakman, they were all caught up in an exciting adventure, converts to a newly revealed gospel expounded by its major prophet right on the premises. No other department of plant pathology anywhere in the world had anything like it. Of course it was exciting.

From the time that graduate education in the department started under Stakman's aegis, in 1913, until shortly before his retirement, in 1953, Stakman was the official advisor to almost all graduate students in the department. The staff member with whom the student was working might direct the student's research, but Stakman was in charge of the student's course work, signed the papers, determined when the student would come up for his preliminary and final oral and written exams and, of course, was chairman of the examination committees. There were a few exceptions to this, but not many. Stakman also was in charge of all research funds administered by the department and so determined within fairly narrow limits what research would be undertaken. This was normal for the times; in addition, not much money was available for optional work.

Insofar as staff salaries were concerned, Stakman was consistent—he kept them as low as possible, probably on the theory that if the crew members were lean and hungry they would row harder. A staff member who was raised in rank from Instructor to Assistant Professor in 1937, with a very commendatory letter from the university president, received an increase of $300 in salary, from $1800 per year to $2100 per year. Stakman himself did fairly well in those days; in 1931–1932 he received a university salary of $5440 plus $1550 from the USDA, the $7000 total being equivalent to about $70,000 in 1981 dollars. In 1940–1941 he received a combined university and USDA salary of $9050, equivalent to about $90,000 in 1981 money. Yet about 1950, according to the head of another department, who was at the meeting of administrators at which money was divided up among the departments, Stakman used his considerable powers of persuasion to argue against raises for most of his staff members.

The field and laboratory work required in the official research projects had to be done (Figure 14), but outside of that, Stakman was not too concerned about how his staff members spent their time as long as they attended seminars. Those with teaching responsibilities taught, without interference but also without much advice or assistance. There never were any discussions or conferences about how teaching might be improved; if the position called for teaching a certain class, the person who held the position taught it as he or she saw fit. Some taught a heavy load; some had little or no teaching responsibility.

Stakman called no staff meetings, except for one each year just before election for officers of the American Phytopathological Society, at which all local members were expected to be present. He did not say, "These are the candidates you must vote for," but by the time the meeting was over (usually about 11:00 p.m.), there was no question in anyone's mind about to whom his vote should go. The candidates backed by Stakman usually won; up until the 1940s, when much of his time became taken up with the Rockefeller Foundation-Mexican Agricultural Program and world agricultural problems, he was a potent manipulator of the affairs of the American Phytopathological Society. Courses

Figure 14. Stakman and J. J. Christensen in the smut laboratory.

in political science were among his favorites in college, and he never lost the urge to influence people and events. Stakman naturally denied, indignantly, during these caucuses, that he was trying to influence in any way the choice of the voters, but the message came through loud and clear. Other than that, he held no staff meetings. There was no need for them. He knew how he wanted his department run, and he ran it that way. One good result of that was that the staff members could get on with the work they were hired to do. He also encouraged work at night and on Saturdays, Sundays, and holidays.

The smooth functioning of the department resulted partly from the efforts of Laura Mae Hamilton, who joined the staff as a clerk in 1921 and retired in 1968. She was an employee of the USDA, not of the University. Whatever her rank and title in the USDA hierarchy, she soon became and throughout her career remained Stakman's personal and private secretary, aide, and amanuensis, giving him invaluable help in preparation of reports, manuscripts, and speeches, and keeping orderly records of all the multitudinous facets of the epidemiology of stem rust. She also served as semiofficial greeter of visiting VIPs and newly arriving graduate students. She and Stakman had a very special relationship, with respect and high regard on both sides (Figure 15).

If Stakman's success as teacher, research leader, and administrator is to be judged by the results—the research done and students trained, which seems a fair basis for evaluation—he rates high marks indeed. Under his headship, the Department of Plant Pathology of the University of Minnesota enjoyed a ranking second to none in the world; unquestionably this was due primarily to Stakman.

Figure 15. Stakman and Laura Mae Hamilton, about 1970.

PART IV
World Agriculturalist

Stakman in Mexico as a member of the Rockefeller Foundation Survey Commission, 1941. (Courtesy, The Rockefeller Foundation)

CHAPTER ELEVEN

The Rockefeller Foundation Agricultural Program in Mexico

The Agricultural Survey Commission

In early 1941, Stakman was asked by the Rockefeller Foundation to serve as one of a group of three experts on a commission that would visit Mexico and report on the advisability of the Foundation's undertaking a program of work to help Mexico improve its agricultural production. The idea for this project had a moderately complex genesis. As described in a letter by Dr. N. E. Borlaug in 1981:[1]

> Vice President Henry Wallace came to Mexico in early December 1940 to represent the U.S. government at the inauguration of the incoming President, Manuel Avila Camacho. Following the ceremonies, he was invited to accompany the incoming Minister of Agriculture, Ing. Marte R. Gómez, and the outgoing President Lázaro Cárdenas, to visit agricultural areas in the Bajio area, which had been opened to cultivation early during the colonial period. At the end of the visit, the Government of Mexico, through Minister of Agriculture Marte R. Gómez, requested technical assistance from the U.S. Government to improve its agriculture. Presumably, Vice President Wallace when he returned to Mexico City from this trip, must have had extensive discussions with Ambassador Josephus Daniels, who was also interested in agriculture.... When Vice President Wallace returned to Washington, war clouds were already on the European and Pacific horizons and there were few possibilities that the U.S. Government could justify assigning scarce scientific work force in the face of these threatening events. Moreover, the U.S. Government had up to that time never participated in foreign technical assistance programs designed primarily to help developing nations. It was presumably for this reason that Vice President Wallace called Dr. Raymond Fosdick and made the suggestion that the Rockefeller Foundation [RF] explore the feasibility of establishing the cooperative Mexican Agricultural Program with RF's support. It is noteworthy that the establishment of the RF-Mexican Cooperative Agricultural Program in 1943 preceded by five years the so-called Marshall Plan for European Recovery Program, which was launched in 1948; and preceded by seven years President Harry Truman's Point 4 Program, which opened the door to foreign technical assistance aimed to assist the developing nations.

The people involved in these early stages of the project were particularly interested in agriculture. Josephus Daniels, who was U.S. ambassador to Mexico at that time, had grown up in North Carolina after the Civil War and had seen tremendous improvements in agriculture there in the early 1900s. A farm demonstration program, about which more will be told later, was partly responsible for these improvements. Daniels was deeply interested in Mexican agriculture and thought that an extension program similar to the one that had done so much good in North Carolina might yield equal benefits in Mexico. He may not have realized, however, that, for an extension program to be effective, the research that develops information for extension people to pass on to farmers or growers must be done with the crops, soils, and practices of the farmers it is designed to help. The general principles can be transferred from Iowa to Minnesota, Mexico, or Timbuctoo, but recommendations for the use of specific varieties or specific practices must be based on research right in the areas where the varieties are to be grown and must fit the practices used.

Henry Wallace, who was vice president of the United States in 1941, had previously served as secretary of agriculture. He came from an established and agriculturally sophisticated family in Iowa; he, like his father, had been the publisher of a farm journal—*Wallace's Farmer*—that was widely read in the area, and the family also grew hybrid corn. Wallace knew good farming practices and the value of research in developing them. He also knew something about farming and agriculture in Mexico, was fluent in Spanish, and was interested in helping modernize agricultural practices and agricultural education there.

Agriculture was not something strange to the Rockefeller Foundation, either. Ever since its founding in 1913, it had been engaged in agricultural activities of one sort or another, although not in the hands-on fashion that was to be used in Mexico and, later, in many other countries. The farm demonstration or extension program that had impressed Josephus Daniels in North Carolina was supported in part by the U.S. Department of Agriculture and in part by the Rockefeller General Education Board. At the risk of straying a bit from the main Stakman theme of the story, this involvement of Rockefeller in agriculture is worth going into briefly. As told by R. B. Fosdick in *The Rockefeller Foundation*:[2]

> Agriculture is one of the oldest interests of the Rockefeller boards. Back in 1903, the newly organized General Education Board undertook a survey of the educational needs of the South. It had no thought of an agricultural program; it was interested only in schools. But the officers had not proceeded far before they realized that the improvement of school systems supported by taxation was inextricably interrelated with the improvement of economic conditions, and since agriculture was the principal source of the South's livelihood, little could be done about the interest in education until something was done about agriculture.
>
> The Board's survey was therefore extended to include this field, and Dr. Buttrick spent almost a year visiting agricultural colleges not only in the South, but in various parts of the country. It was while he was in Texas that he learned of the work of Seaman A. Knapp.
>
> Texas at this time was the chief battleground in the fight against the boll weevil, a pest which had recently crossed over from Mexico and was devastating the cotton fields. Dr. Knapp, a staff man of the United States Department of Agriculture, had established a demonstration farm in the

infested region to show farmers how, despite the boll weevil, they could produce satisfactory crops. By such measures as careful seed selection, deep plowing in the autumn, wide spacing of plants, intensive cultivation, systematic fertilization, and rotation of crops, Dr. Knapp had stepped up the normal yield in his demonstration farm substantially, and farmers who went back to their fields and applied his methods were increasing their yields too. Dr. Buttrick was deeply impressed. "If the demonstration method paid in dealing with a pest-ridden farm, was there not every reason to suppose that it would pay still more handsomely where no handicap existed?" he reasoned. "In Dr. Knapp's farm demonstration work, limited at that time to combatting the boll weevil, the General Education Board found the answer to its search for a method of delivering the existing knowledge of effective agricultural processes to present farmers."

A co-operative arrangement was worked out. The United States Department of Agriculture agreed to carry on the demonstrations in weevil-infested states and the General Education Board agreed to provide funds for extending the work to other states, both programs to be under Dr. Knapp's direction. This joint effort, launched in 1906 in a few rural centers in six states, rapidly multiplied until the entire South was dotted with nearly one hundred thousand demonstration farms. Later, the program was extended to Maine and New Hampshire. Each demonstration farm served its surrounding territory as a practical seminar in scientific agriculture, showing hundreds of farmers how to engineer a crop to a successful yield. From cotton, the program was extended to embrace corn and other crops, and it proved an important part in the movement for crop diversification.

By 1914, the General Education Board felt that the practical value of the work had been established and its continuance could be left to others. The Board had invested close to a million dollars in the farm demonstrations, the United States Department of Agriculture had invested double that amount, and, significantly, the Southern people themselves had put more than a million dollars into the effort. The demonstration farms paved the way for the development a few years later of the nation-wide agricultural extension service, supported by public funds and administered jointly by the Department of Agriculture and the land-grant colleges.

Probably few people today are aware of the fact that the wide-ranging federal-state-county extension service, which has played so large a part in the development of scientific agriculture in the United States, was in part the product of the Rockefeller General Education Board. Probably few people way back then were aware of it either, since that was the way the Board, and subsequently the Foundation, operated.

To return to Stakman and the Mexican Agricultural Program—the Rockefeller Foundation had long had medical teams in foreign lands, including Latin America, and it was aware of the relationship between health and diet and therefore the importance to health of farming. Dr. Warren Weaver, then Deputy Director of the National Science Division of the Foundation, was interested in the possibility of sponsoring a program for agricultural development, if it seemed like a good gamble, and Vice President Wallace suggested that the Foundation consider supporting such a program in Mexico. To get an objective view of this possibility, the Foundation decided to send a survey commission to Mexico to look the situation over and report on it. For this commission they selected three men, Dr. Richard Bradfield, Head of Agronomy and Soils at Cornell University; Dr. Paul Mangelsdorf, Professor of Plant Genetics and Economic Botany at Harvard University; and Dr. Stakman. The commission members were to travel through Mexico (a Dodge Carryall was furnished), observe farms and farming, agricultural research, agricultural schools and teaching, and agricultural extension and administration, and, at the end of the trip, write a report.

Three better men for the task probably could not have been found. Each was outstanding in his own field but knew much about agriculture in general, both in theory and in practice. If they were not exactly dirt farmers, they were personally familiar with field work and with the mechanics of crop production. All three were educated, thoughtful men of large vision. Among them they had had 75 years of experience in research, teaching, and administration in subjects basic to modern agriculture. They were at home with persons of all ranks, from peasants to presidents. And, very important for the success of the survey, they were congenial companions and enjoyed one another's company. On some subjects they might have different and fairly strong opinions, but they could argue out their differences without rancor or antagonism. They remained friends throughout their lives.

Stakman had been to northern Mexico many times before, in his rust surveys from 1917 on, and so he skipped the first part of the tour but joined the others shortly. They traveled the highways and byways of Mexico for two months, covering 5,000 miles from the border of Texas to the border of Guatemala, by station wagon, flatbed truck, horse-drawn or mule-drawn buckboard, train, plane, and, occasionally, in areas not accessible by roads, on horseback or muleback.

They visited 114 towns and cities in 16 of the 33 states of Mexico. They observed, talked with, and listened to peasants who used wooden plows drawn by bullocks and who harvested wheat with a sickle and threshed it by treading and winnowing, as in biblical times; to boys and girls and teachers at rural schools; to students and professors at agricultural schools (Stakman had been visiting some of the agricultural schools in northern Mexico, on his rust surveys, since 1928 or before); to research and extension personnel; and to state and federal government officials. They gathered statistics on crop production from official and unofficial sources. At times it must have been wearying. They got shook up on country roads and stuck in mudholes in tropical downpours, and they must have experienced some primitive accommodations, with fleas and dysentery, but these hazards they accepted without comment. In all the photographs of the trip, the three men, whether standing in a flatbed truck while the driver shoved something under a rear wheel to extricate the truck from a mudhole or sitting on muleback like three modern Don Quixotes, are wearing their regular business suits, with white shirts and ties. Stakman, then 56 years old, was a pretty tough and hardened campaigner. Even in his late 70s he could endure wearying plane trips followed by exhausting days in the field under the hot sun, with much younger men, and would ask no quarter; at sundown or after he was as fresh and vigorous, physically and mentally, as he had been early in the morning. He had seemingly inexhaustible resources upon which to draw. At times, when he returned to St. Paul from one of these strenuous trips, he would spend a week or more in bed with an ailment that he diagnosed as flu but that probably was mostly exhaustion.

When the survey was finished, the three men holed up and wrote a report of 62 single-spaced typewritten pages, illustrated by photographs taken by Mangelsdorf, and submitted it to the Rockefeller Foundation. The report[3] begins:

It is the unanimous opinion of the Rockefeller Agricultural Survey Commission to Mexico that there is urgent need for improving agricultural conditions and practices, that the most important problems are obvious, that the time is propitious, and that there now is potential and partially functioning talent to justify the opinion that substantial improvement could be accomplished with even a moderate amount of help from an outside and independent agency, such as the Rockefeller Foundation.

Approval by the Foundation officers followed shortly, but because of a commitment to unfinished business by the man chosen to head it, the program did not get under way until 1943. No one then could have foreseen that this modest and somewhat tentative venture would evolve into a worldwide network of research centers that still have a tremendous effect not only on the developing countries but on all countries. Stakman must be given at least a small measure of credit for this. Again the fates had conspired to bring about a combination of circumstances in which Stakman could apply his inexhaustible energy, his judgment, and his vision to the good of his fellow humans.

The Status of Education, Research, and Extension in Agriculture in the United States and in Mexico in 1941

Education in agriculture as a separate discipline or course of study was begun in some eastern colleges of the United States before 1850, more in response to demands by farmers and influential farm groups than from any conviction among college administrators or educators that a curriculum in agriculture would be a desirable innovation. The early attempts to teach agricultural subjects probably didn't amount to much, because not many facts and principles were known, but at least it was a beginning. In 1862, Congress passed the Morrill Act, allotting to each state parcels of land (30,000 acres per parcel) equal to the number of congressional representatives and senators from that state. The income from this land was to be used to establish and support education in agriculture and the mechanic arts in colleges and universities. This foresighted measure was partly responsible for the later rise of the land-grant colleges (now mostly universities), but at the time it did not have much immediate effect on agricultural education in at least some of the states. The situation at the University of Minnesota in the mid 1870s is indicated by the following quotation from Folwell's *History of Minnesota*[4] (Folwell was president of the university at the time, and even when he wrote the history, some 40–50 years later, he still had a somewhat unenthusiastic view of things agricultural, at least so far as the university was concerned):

In the year 1874 the regents thought the time had come to fill the chair of agriculture with a well-educated and qualified expert. They called into service Charles Y. Lacy, an agricultural alumnus of Cornell University, twenty-four years of age, and put him in charge of instruction and of the experimental farm. They were just completing the first agricultural building on the campus, a building of moderate size but well proportioned and well built, with one wing for a chemical laboratory and another for a plant house, both well equipped. It was hoped that this physical demonstration would call in students to pursue agricultural studies in good faith. There were plenty of men students who would gladly work on the farm for customary wages, but hardly one of them would suffer himself to be enrolled as a student of agriculture. In his first year Lacy had no student of college rank. In the next, 1875-76, there were three students, one of whom dropped out at the end of the winter term. In the third year there were two students, one of whom escaped.

Something more than a well-proportioned building containing a chemical laboratory and a plant house, both well equipped, was required to initiate an effective program in agricultural education and research. This had to await someone more perceptive of the needs of agriculture and more interested in filling them than was President Folwell, and, indeed, most of his successors. In 1887, the Hatch Act was passed, providing federal funds to establish, promote, and help support agricultural experiment stations. Many states already had such experiment stations, and most states soon would have them. By the time Stakman began graduate work on the St. Paul campus, in 1909, there were several hundred students in the College of Agriculture, Forestry and Home Economics.

The College of Agriculture in Minnesota and those in most other states were long referred to disparagingly by the faculty and students in the arts colleges as "cow colleges." The general attitude was that those in the arts colleges were the elite, the intellectuals, and those in the cow colleges were clods and rubes, shy on social graces and not really of university caliber. This would not be worth mentioning were it not for the fact that this view was shared by many university administrators, even in states where agriculture constituted the major source of wealth that supported the university. An honors graduate on the "Farm Campus" of the University of Minnesota, for example, received a diploma inscribed "With Distinction" or "With High Distinction," not "Cum Laude" or "Magna Cum Laude," as in the arts colleges, implying that although some of the Farm Campus boys might be smart, they should not be accorded the academic recognition reserved for polite learning and for scholars. Some university administrators were slow to grasp the significance of agriculture in the development of civilization and in the development of the very societies in which they lived and worked and on whom they depended for their support. In most states, the state colleges, where agricultural education was based, developed separately from the universities and thus avoided being under the domination of administrators and governing boards unsympathetic to or even antagonistic to education and research in agriculture.

Following World War I, teaching, research, and extension in the agricultural sciences developed rapidly. The federal government's involvement in agriculture ranged from influencing the agricultural curriculums in high school through graduate or postgraduate work in colleges and universities, to maintaining experiment stations and the U.S. Department of Agriculture. The experiment stations, supported in part by federal funds and staffed in part by persons on federal appointments, became an integral part of the overall effort. It was not at all uncommon, as in Stakman's case, for a member of the university staff to have a part-time federal appointment, and some graduate students were on assistantships paid for by federal funds.

County agricultural agents were university educated and maintained close connections with the university. Their offices were furnished by the counties, but they were paid by federal funds. Many of these county agents (now extension specialists) developed a very high degree of expertise in their special fields and in agriculture in general, and they were looked up to as important members of their communities. In agricultural research, teaching, and extension in general, there developed a tradition of service and dedication, of obligation to the public, of pride in their work. Agribusiness firms gradually became increasingly involved in many aspects of research and extension, and many firms not only collaborated closely with university workers but also contributed generously to the support of the work. The standoffishness and mutual suspicion that had existed between university and industry personnel at the time of the barberry eradication campaign in 1918 was disappearing in the 1930s, as both groups began to recognize that they were working for a common goal of increased productivity and that it made sense to work together rather than be adversaries. Many of the industrial fellowships were continued for decades and still constitute an important part of graduate education and research in agriculture.

As one example of how the extension system presently functions, suppose that during the growing season, a farmer notices what seems to be an unusually large number of sickly or dying plants where experience would indicate healthy plants should be. That farmer is likely to call the county agent for an opinion. The county agent may in turn contact specialists at the university—agronomists, horticulturists, plant pathologists, entomologists, or others. If the problem appears to be of more than minor significance, the local newspaper and radio station are likely to publicize it. If, after examination, the problem turns out to be something new, one of the specialists in whose field the problem lies may undertake work on it. If a number of growers are affected, they may get in contact with their legislator, who in turn calls the experiment station director to see what can be done. If more research is needed than the available work force can furnish (most research people in most experiment stations are not sitting around waiting for new problems to appear), funds may be made available quickly to get work started. That is, although much of the experiment station research is of an ongoing, long-time nature to help take care of the future, the system provides enough flexibility so that money and resources in the work force can be allocated to immediate problems too. It is, in general, a very effective system, which has had a tremendous influence on agriculture in the country. Having evolved over a period of a century, it is in reality not one uniform system throughout the land but a number of different systems, each supposedly suitable for the region in which it developed and for the growers it serves. These systems involve people and their interrelationships, which means that they involve psychology, habits, traditions, and prejudices. The principle can be transferred to another country; the specific system cannot.

Mexico, in 1941, had almost none of the ingredients of this system. At that time, there were three agricultural schools in Mexico, one of them private. The students in these schools were taught theory, not practice; they had very little laboratory and no field work. Field work was for peons, not professors. The Survey Commission stated that:

the most acute and immediate problems, in approximate order of importance, seem to be the improvement of soil management and tillage practices; the introduction, or breeding of better-adapted, high-yielding and higher-quality crop varieties; more rational and effective control of plant diseases and insect pests; and the introduction or development of better breeds of domestic animals and poultry, as well as better feeding methods and disease control.

The commission therefore recommended that the Rockefeller Foundation initially send to Mexico "an agronomist and soils expert, a plant breeder, an expert in crop protection, and an animal husbandman as a working commission to cooperate with the Mexican Ministry in its agricultural improvement program." In 1967, the three members of the commission published a book, *Campaigns against Hunger*,[5] describing the work in Mexico and, later, in other parts of the world. Of the original project in Mexico, they said:

the ultimate aim was to help Mexico toward independence in agricultural production, in agricultural science, and in agricultural education. The design was joint participation, not preachment; the intent was to work with Mexicans in doing the things that needed to be done, not merely tell them how things should be done. As independence would depend upon the number and quality of independent Mexican scientists, the Commission recommended "that provision be made for a special type of fellowship for outstanding investigators and teachers."

They also emphasized the importance of developing a strong extension program to get the research results to the farmers. The Foundation knew, from long experience, that one of its most important functions was to educate those it was trying to aid; the General Education Board, which preceded the Foundation, and the International Education Board, both of which eventually were combined with the Foundation, were established for the purpose of aiding education.

As head of the Mexican Agricultural Program, the Foundation chose Dr. J. G. Harrar (Figure 16), who had received his Ph.D. in plant pathology in Stakman's department in 1935. Harrar had then served as head of the Biology Department (and as track coach) at Virginia Polytechnic Institute and as head of the Department of Plant Pathology at Washington State College (now University) at Pullman. Before coming to Minnesota for his doctor's degree, he had taught for four years at the University of Puerto Rico and was very competent in Spanish. Again the Foundation made a most fortunate choice. As Dr. R. B. Fosdick states in his history of the Rockefeller Foundation,[6]

Dr. J. George Harrar ... was engaged to head the work as the Foundation's field director for agriculture; and the brilliant success which this new type of activity has met is in no small degree due to his character, judgment, and distinguished capacity for leadership.

Harrar later became director of agriculture, then vice president, then president of the Rockefeller Foundation. Still later, he was a prime mover in the organizing of a network of agricultural research centers in developing countries throughout the world, all of them more or less lineal descendants of this Mexican venture that he was chosen to head and that he headed so effectively.

Figure 16. Stakman and J. G. Harrar in the field in Mexico, about 1945.

Among the important diseases of field crops in Mexico were the rusts of wheat. Their status in 1944 is indicated by the following quote from Stakman and Harrar (the article was written in 1944, a year after the Mexican Agricultural Program got under way):[7]

> The most typically epidemic diseases of crop plants are, however, the rusts of wheat. Orange leaf rust (*Puccinia rubigo-vera tritici*), yellow stripe rust (*P. glumarum*) and stemrust (*P. graminis*) are generally distributed, but the destructiveness of each varies greatly with the region and the season. Most of the varieties commonly grown are susceptible to all three rusts; consequently they can propagate and spread rapidly under favorable weather conditions. Moreover, some wheat is grown throughout the year. Although most of it is grown as a winter crop, some of it is grown in the summer also. Even where there is no summer crop, there are likely to be some volunteer plants on irrigated land, and the rusts may therefore survive in the uredial stage throughout the year in enough places and in sufficient quantity to become locally or regionally epidemic whenever weather conditions favor rust development for sufficient periods of time. Nevertheless, avoidance of out-of-season sowings of wheat would help measurably in reducing rust losses.
>
> Stem rust is not only the most destructive disease of wheat in Mexico, but is considered by informed Mexican agriculturists as one of the important problems in crop production. Not only does it sometimes cause devastating epidemics but it also limits the areas in which wheat can be grown profitably. As the Mexican government desires an expansion of wheat growing, control of stem rust is their most pressing plant disease problem. And its importance is not limited to Mexico alone.
>
> Actually, the control of stem rust of wheat in Mexico is almost as important to the United States and Canada as to Mexico herself. It is an international problem. The general features of the rust development in Mexico are known as a result of long-time studies made cooperatively by the Mexican and United States Departments of Agriculture. It is known that there may be a seasonal interchange of rust between the two countries. Rust from the United States may be blown into northern Mexico in the fall and rust from northern Mexico may be blown into the United States in late winter and early spring....
>
> The Mexican Department of Agriculture in cooperation with the United States Department of Agriculture, the Rockefeller Foundation, and the Rust Prevention Association of Minneapolis, Minnesota, is now making extensive varietal tests of wheat in well-selected localities in each of the major wheat growing regions. As a result of this work, it seems possible that resistant varieties can be selected from among those already available, even though different varieties probably will have to be selected for different regions or possibly for areas within regions. Prospects are, however, that fairly rapid progress can be made.

Fairly rapid progress *was* made. The phenomenal success of the Mexican Agricultural Program and the first of its offshoots—in Colombia, Equador, and Chile in South America and India in the Far East—is described in all its fascinating detail by Stakman, Bradfield, and Mangelsdorf in *Campaigns against Hunger*.

After his retirement as head of the Department of Plant Pathology at the

University of Minnesota in June 1953, Stakman continued as a consultant to the Rockefeller Foundation programs in agriculture and spent much of his time in the Office of Special Studies, the headquarters of the Mexican Agricultural Program. There he kept his finger on the pulse of the whole organization and used his influence where he thought it would further the cause of the program. One of his concerns, and one to which he devoted a good deal of time and thought over the years, was the establishment of a graduate school at the National Agricultural School at Chapingo, Mexico. This was a long, slow project beset with many difficulties, especially as some of the old-timers at the school professed to see in this idea a plot by "Yanqui" imperialists to take over higher education in Mexico, but Stakman persisted and lived to see the graduate school at Chapingo become a reality. Beginning in 1958, Master of Science degrees were awarded there, and in 1973, the Ph.D. degree was added. Stakman was present at the gala celebration when the new reading room, named for him, was dedicated. He also helped organize the Mexican Society of Plant Pathology and contributed to the planning of its first two national meetings. His last visit to Mexico was in May 1977, when he was honored by the National Institute of Agricultural Research (INIA) and the International Center of Corn and Wheat Improvement (CIMMYT) at Ciudad Obregon, Sonora, and by the graduate school at Chapingo. It must have been rewarding.

Some Accomplishments of the Mexican Agricultural Program and Offshoots from It in Stakman's Time

Although the accomplishments are told so well in *Campaigns against Hunger*, a few comments may add some flavor and substance to the Stakman story, since he was so closely associated with so much of what went on there.

Very soon after his arrival in 1943, Harrar established close and cordial working relations with men in various agencies and bureaus of the Mexican Ministry of Agriculture, established test plots for field and experimental work, and began to build a staff of young, able, vigorous, and dedicated men, who developed within a few years into what must have been one of the most outstanding agricultural research and production teams ever assembled anywhere.

Probably at Stakman's suggestion, Harrar chose Norman E. Borlaug to head the wheat improvement aspect of the program. Borlaug, a farm boy from Cresco, Iowa, had graduated in forestry from the University of Minnesota in 1937 and had stayed on to do graduate work in plant pathology. He received the M.S. in 1941 and the Ph.D. in 1942. He and Stakman became and remained very close friends and mutual admirers (Figure 17); among other things, they shared similar views about their work and goals—that those were to be pursued with total dedication and all resources, 24 hours a day, 365 days a year. At the time Harrar invited him to Mexico, Borlaug was working for the DuPont Chemical Company. He stayed with the RF-Mexico program until 1960. In 1963, he became director of the worldwide wheat program at CIMMYT. In 1970, Borlaug was awarded the Nobel Peace Prize for his contribution to improved food production throughout the world.

By 1948, Mexico was self-sufficient in corn, for the first time in 35 years. By 1956, it was self-sufficient in wheat, with new varieties that were resistant to stem and leaf rust, had high milling and baking quality, and were widely adapted.

Proper use of fertilizers and soil management techniques developed by soils researchers played a big part in the increased yields, and the importance of these techniques was becoming recognized by progressive farmers. The yield of beans was increasing, not dramatically, but consistently. Potatoes, which had never amounted to much as a commercial crop because of the ravages of diseases, were now being grown successfully by small farmers for food and by large farmers for sale, thanks to newly developed varieties resistant to late blight and free of viruses. New forage crops were being grown; production of milk and eggs was increasing.

Hand in hand with all this went the training of interns from the Mexican schools of agriculture, who toiled and sweated in the fields right along with Harrar's men and who now not only wore field clothes but were proud of it. A tangible spirit of pride, accomplishment, and adventure pervaded the whole organization. A few excerpts from *Campaigns against Hunger*[8] show this.

> The Foundation scientists in Mexico took pride in their experiments and were eager to show them to visitors. Exhibits were set up and field trips arranged. One corn farmer near Texcoco, not far from the experiment station at Chapingo, was interested in obtaining seed of one of the strains under test. The supply was very limited, but Wellhausen agreed to provide him with some seed if he would plant it, fertilize it, and manage it throughout the season as directed. The farmer, in turn, agreed to permit the use of this planting as a demonstration of what could be done on his farm

Figure 17. Stakman and Norman Borlaug during a visit by Borlaug to the University of Minnesota in 1970.

when the best management practices known at that time were combined. He carried through splendidly and was rewarded with a fabulous crop. A field day was arranged, and the President of the Republic, Miguel Alemán, attended with many of his high officials. The press naturally was on hand. Farmers had never seen such corn! When harvested, it yielded over 125 bushels an acre—one of the highest yields ever reported in the Valley of Mexico. The Office of Special Studies was soon deluged with inquiries from farmers who had read newspaper accounts and wanted more information about how they too could grow such bumper crops.

In the spring of 1948 Sr. Aureliano Campoy, a large wheat farmer from Sonora who had been cooperating with Borlaug in his wheat experiments in the Yaqui Valley, came to Mexico City to visit a nephew who was editor of El Universal, one of the city's leading daily papers. After describing Borlaug's work and what it had meant to him personally as well as to the other farmers in Sonora, Sr. Campoy urged his nephew to visit nearby Chapingo to see the experiments in progress. A field trip was arranged for the editor and several members of his staff. Young Mexican scientists were on hand to explain the different experiments with the various crops. The visit lasted almost all day; feature writers asked hundreds of questions, staff photographers took scores of pictures. The feature article in El Universal the next morning was devoted to what the staff had seen and learned at Chapingo. And because the paper was widely read throughout Mexico, editors of dozens of local papers saw the article and republished selections from it.

This was only about four years after the program had gotten under way. The progress continued on all fronts, and word of its success soon spread far beyond the borders of Mexico. In reponse to requests, the Foundation began similar programs in Colombia in 1950, in Chile in 1955, and in India in 1956. It also established lesser missions in Ecuador and Peru. In all of these, Stakman had a deep interest and some direct input.

CHAPTER TWELVE

Agricultural Research Centers Established Around the World

In 1962, after some years of preliminary study, the Rockefeller and Ford Foundations jointly established the International Rice Research Institute (IRRI), near Manila, Philippines, supported by them with some help from USAID. The second center of international agricultural research, CIMMYT (Centro Internacional de Mejoramiento de Mais y Trigo—International Center of Corn and Wheat Improvement) was established in 1966 at El Batán, near Texcoco, Mexico, and near the National Agricultural School at Chapingo, which is also the location of the headquarters of INIA, the Instituto Nacional de Investigaciones Agricolas—Mexico's National Institute of Agricultural Research. CIMMYT, like IRRI, was a product of Rockefeller and Ford Foundation vision, but it was supported by contributions from many sources and had an international board of directors. Although Stakman had relatively little input into the development of these research centers, they were more or less a natural outgrowth of the agricultural programs in Mexico and other Latin American countries that he was directly involved in founding and for which he was partly responsible.

As of 1981, 13 such international agricultural research centers existed, under the general aegis of the Consultative Group on International Agricultural Research (CGIAR), based in Washington, DC. The centers, located in developing countries around the world, are staffed by trained research people from many countries and have international boards of directors. Some excerpts from the 1980 publication, "Consultative Group on International Agricultural Research,"[1] tell how these centers came about and what they are intended to be.

> At four meetings during 1969 and 1970, the leaders of the major national and international funding agencies reviewed the opportunities for cooperation in increasing food production in developing countries. In October 1969, the president of the World Bank proposed to the United Nations Development Program [UNDP] and the Food and Agriculture Organization [FAO] that the three institutions jointly organize long-term support for an expanded international agricultural research system.
>
> The result of these initiatives was the establishment in 1971 of the Consultative Group on International Agricultural Research .... Like the international centers, the Consultative Group is an unconventional

institution. It operates without any legal charter, written rules, protocols, or by-laws, entirely by common consent, shared interest, and goodwill of its members. Meetings [are] held once or twice yearly to consider program and budget proposals, policy issues, and other matters from the centers.... The CGIAR tries to maintain a delicate balance between informality and the need for accountability to assure the efficient and effective use of large amounts of money in a flexible, decentralized system.

The report also describes the international research centers:

The international research centers, themselves, constitute an important new approach to increasing food production in developing countries. Independent and autonomous, responsible to boards of trustees selected from the world's distinguished agricultural scientists and policymakers, the centers can pursue long-range research goals with fewer of the staffing and funding constraints and the political pressures common in national institutions. Widely recognized for scientific excellence and worthy purpose, the centers attract talented, dedicated scientists from all over the world interested in finding practical solutions to the world's food problems. The centers cut across traditional academic lines to form multidisciplinary teams of diverse specialists for the improvement of major crops and farming systems, supported by research sources not available in most national programs. With worldwide germplasm collections and international networks for testing and adapting new materials, the centers can provide a rapid flow of technologies to national programs.

The international centers also serve as unique training institutions to strengthen the capacities of developing countries to carry out their own research, as well as to collaborate with the centers in developing, testing, and adapting new technologies. Most of the several hundred trainees who come to the centers each year spend 3 to 12 months working under the guidance of senior staff scientists and training specialists in the fields and laboratories— a novel experience for many agricultural graduates from developing countries, where academic training often does not include practical experience in actual farming or production research. The purpose of the in-service training is to produce the well-prepared, dedicated researchers and other specialists sorely needed in the developing countries. The centers also afford research opportunities to M.S. and Ph.D. candidates, postdoctoral fellows, and visiting scientists whose projects are relevant to the centers' primary mission.

Support for these centers comes not only from the World Bank, the UNDP, and the FAO of the United Nations, but also from 35 countries, international and regional agencies, and private foundations. The 1980 budget was $120 million.

Basically, the plan of operation as described above is very similar to that developed in the Mexican Agricultural Program. None of those who conceived, gestated, and nourished the Mexican Agricultural Program in the 1940s could have imagined that it would grow into such a mighty engine for worldwide agricultural improvement. Some of these centers already existed in the 1970s, and others were in the planning stages, so Stakman knew about them, and in his more optimistic moments he must have reflected on their possible contributions to a better world. He was very much aware of the race between increasing food and increasing population, as exemplified by Mexico's increase from about 19 million people in 1940 to 35 million in 1960 (and more than 70 million in 1980), but he remained hopeful that the runaway increase in population in the developing countries could be controlled by some means other than plague, war, or famine. Whether it can be controlled in time, no one knows. Certainly much greater resources are now being mustered to attempt to improve the lot of people, especially in the developing countries, than anyone could have imagined in 1941 when the Rockefeller Foundation sent the survey commission to Mexico to look at the possibility of increasing agricultural production there. Perhaps it is unrealistic to think that the forces for good might win out, over a great portion of the world, and that there will come a time of peace and plenty such as all people of reason and compassion hope for, instead of the dreadful alternatives. But whatever the final outcome, today more citizens in the developing countries are eating better and walking more upright and with more control over their own destinies than they were a short generation ago. They have a chance. For whatever he contributed to this, Stakman deserves a measure of thanks; nothing would have pleased him more than to know that he had contributed significantly to the welfare of mankind.

PART V

Public Spokesman/
Private Citizen

Christmas 1947 at the Stakman home. C. M. Christensen (standing),
J. G. Harrar, and Stakman.

CHAPTER THIRTEEN

Speeches and Writings

Although Stakman grew up in a small midwestern town, by the time he had graduated from college, in 1906, his outlook already had an international cast; he was familiar with the language, customs, and ways of thinking of ancient Rome and of contemporary Germany. Other lands, other cultures, had an attraction for him; he wanted to get to know the people as fellow human beings, to share their hopes and fears, and especially to understand them.

Very early in his work he developed a broad view, and this was strengthened as he became increasingly involved in plant pathological problems and agricultural activities of international scope—making stem rust surveys, for which he repeatedly traveled from Canada to Central Mexico; surveying barberries and barberry eradication in Europe; teaching the graduate students whom he attracted from a dozen or more countries from around the world; and being involved in the Rockefeller Foundation-Mexican Department of Agriculture Program and the larger international programs that followed. From about 1940 on, his views and opinions were increasingly sought on problems of general import, and he became to some extent a spokesman for plant pathology, for research in agriculture (and research in general), for the place of science in society, and for international cooperation as an essential ingredient of a rational world. He became an elder statesman of science; his award from the Cosmos Club in Washington in 1964 was, in fact, entitled "Statesman of Science."

His activities on the lecture platform reached a peak in 1949. In that year, he gave 32 speeches—in Minneapolis, St. Paul, Willmar, and Morris in Minnesota; Vancouver and Winnipeg in Canada; Philadelphia, Pennsylvania; Milwaukee, Wisconsin; Gainesville, Florida; Lake Success, New York; and Mexico City, Mexico. Several of the talks dealt with various aspects of plant pathology and stem rust; four of them were on science and its sphere of influence; four were on problems of human subsistence. Stakman was an experienced and effective public speaker, and, like most public speakers in demand, he had a message, or messages, which is evident in the excerpts below. America is the land, among other things, of the inspirational public speaker, in which personal magnetism, stage presence, and delivery are likely to be valued above content. After all, the content can be given much more effectively and cheaply in writing, with the added advantage that the reader can digest it in comfort and alone, and can reflect upon what he has read if reflection is called for. We like to be exhorted, and Stakman was a good exhorter, but he also had some very solid material to offer to his listeners and readers. He accepted remuneration for expenses incurred, but consistently refused fees, even from those well able to pay and accustomed to paying. The selections in this chapter are a representative sample of his contributions from 1947 on.

Stakman was a member of the American Philosophical Society, probably the only plant pathologist and one of the few biologists ever accorded the honor of membership in that select group. Benjamin Franklin, who founded the society in 1743, would have enjoyed Stakman's company; both were great talkers and great doers.

In a talk in 1947,[1] Stakman told his audience:

> Plant diseases, insect pests, and unfavorable weather are among man's worst enemies in his attempts to feed and clothe himself, because man is basically dependent for his subsistence on plants and plant products. After all, plants are the only primary producers. Animals, important as they are, are essentially only transformers of plant products, not primary producers. They themselves are directly or indirectly dependent on plants for their subsistence. And ever since man left the Garden of Eden he has been confronted at some times and in some places with the specter of famine resulting from crop failures due to droughts, floods, frosts, and the ravages of insect pests and plant diseases.

The main body of this talk dealt with epidemiology of stem rust, with cereal smuts, and late blight of potatoes; it ended with:

> Every thinking person must hope that the ambition to expand the boundaries of knowledge will always burn within many human breasts, whether the motivation be pure intellectual curiosity or the desire to render social service. Motives probably often are mixed. I cannot refrain from saying that an infinity of problems in the field of plant pathology are insistently fascinating and challenging, even though their solution would have no foreseeable practical value. And the solution would tax the intelligence and inventiveness of the best minds in the various scientific guilds within the fields of biology, chemistry, and physics. This is a tribute that I want to pay to the science of plant pathology, especially since it has seemed necessary to emphasize in this paper the vital necessity of international cooperation in helping, through controlling plant diseases, to remove at least some of the fear of want among human beings, wherever they may be.

Stakman always ranked plant pathology right up among the top learned professions, in its intellectual demands, its contributions to society, and its spiritual if not its financial rewards. He was able to convey this sense of importance to his listeners, whether they were farmers, undergraduate or graduate students, businessmen, politicians, or philosophers. When an ordinary man said some of the things that Stakman said in his speeches, they sounded ordinary and mundane; when Stakman said them, they sounded like profound truths and had the force of revelation.

In 1950, Stakman participated in a Symposium on Intellectual Freedom sponsored by the American Philosophical Society.[2] The stimulus for this symposium was the publicity given to what purported to be a revolutionary discovery, in the USSR, of a new principle for the production of new and improved varieties of economic plants and animals. It was known as "Lysenkoism," from Lysenko, its foremost promoter. (Another Soviet scientist, Michurin, was mostly responsible for developing it, but Lysenko publicized it.) According to Lysenko and his followers, if plants or animals were exposed to the proper environmental or physiological conditions, the desired new characters would appear more or less automatically and then would become fixed and capable of being passed on to descendants. By this means, they had even transformed plants of one genus into those of another. It was a new method of producing superior plants and animals, much better than the slow and chancy methods used by the plant and animal breeders of the decadent capitalist countries. In his talk, Stakman quoted Lysenko:

> We, the representatives of the Soviet Michurinist trend, contend that inheritance of characters acquired by plants and animals in the process of their development is possible and necessary. . . . Michurin's teaching shows every biologist the way to regulating the nature of plant and animal organisms, of altering them in a direction required for practical purposes by regulating the conditions of life, i.e., by physiological means.

This was supported by the Communist Party, the Soviet government, and by Stalin himself. Stakman again quoted Lysenko:

> The Presidium of the Academy of Sciences of the USSR obligates the Department of Biological Sciences, biologists, and all natural science research workers, working in the Academy of Sciences, to make a basic reorganization of their work, to occupy a leading place in the struggle against idealistic, reactionary teachings in science, against servility to foreign pseudo-science.

That was pretty strong stuff. The research approach in the USSR was determined by the Communist Party, and those who did not conform to the party line either made a "basic reorganization of their work" or were ousted, some presumably to Siberia.

Lysenkoism did not shake the foundations of genetics in the West, but some people evidently were taken in by it, including the supposedly sophisticated science writer for the prestigious *New York Sunday Times*. Most legislators, who control funds for research, have little background in biological sciences, certainly none in the intricacies of modern genetics, and if they heard about the wonders of this new Lysenkoism, they might question whether some of the funds they were providing were being wasted on an outmoded approach in spite of all the tangible benefits that had accrued. In any case, someone in authority had to expose Lysenkoism for what it was. Even more important was to point out the danger of attempting to make research conform to authoritarian dogma. The symposium on intellectual freedom was a good place to do it, and Stakman was a good man for the job. Lysenko and Lysenkoism are ancient history now; few students have even heard of it. After a time it was even discredited in the USSR, where they now grow wheats derived from those produced by Borlaug and his co-workers in CIMMYT, that bastion of servile and decadent capitalism. Only a few portions of Stakman's relatively long paper will be quoted here. He started out,

> Science has been extricated from the entanglements of magic and it appeared to have been largely emancipated from ecclesiastic and political dogmas also. So great had been the progress that [F. Sherwood] Taylor wrote in *The March of Mind* . . ., "The conflict of Science and Authority ended with the Evolutionary Controversy in the seventies. Since that time

the scientific method of objective observation, experiment and statistical examination has invaded every department of knowledge. Whatever has to be investigated today, is investigated by the scientific method, no matter whether the subject is the mental processes of animals, the optimum conditions for growing beetroot, or the type of chocolate preferred by insurance clerks. A century ago, the approach to these subjects, insofar as they would have commended themselves as matter for investigation at all, would have been the construction of a theory on more or less a priori grounds supported by a few well-known facts. Today the tendency is to discard theory and to collect and classify great numbers of instances and obtain from them some statistical rules."

Stakman commented,

The misunderstanding or misrepresentation of the miscalled "reactionary, idealistic" genetics and the glorification of the so-called "progressive, materialistic Michurinist trend" might be expected to emanate from impatient men primarily interested in quickly obtaining results of practical value. But the pronouncements are made by the Presidium of the Academy of Sciences of the USSR, which has wide powers in determining policies and administering funds for research. Moreover, the general condemnation of non-conformists is followed by orders for removal or demotion of individuals and abolition of laboratories because of their alleged "anti-scientific" position....

Had this happened a century ago, a modern biologist probably would say, "It could not happen now." But it has happened now, and in Russia, where biology and its applications were making commendable progress.

Americans are likely to say, "It could not happen here." But it has happened here, and not long ago. The controversy among Russian scientists was not whether evolution is a fact but rather about the facts of evolution, about the way evolution comes about. The authoritarian settlement seems amazing in this relatively enlightened age.

We should recall, however, that there raged in the United States of America, not longer than a score of years ago, a controversy as to whether evolution was even a fact. Fundamentalist non-scientists attempted to discredit scientists by debate, ridicule, or mere volume of verbiage; and, unable to prevail by forensics, the anti-evolutionists attempted to prevail by law. Bills to prohibit or restrict the teaching of evolution were introduced into seventeen state legislatures and were enacted into law in two. Though the laws probably did little harm, the world was treated to the grotesque spectacle of the "Dayton monkey trial." It would be unwise to forget that authoritarian methods were used successfully by scientifically ignorant but strongly organized groups to suppress or restrict the teaching of scientific facts. And this was only one example of powerful attempts to impose authoritarianism in other fields—history, economics, and sociology. Unorthodoxy becomes un-Americanism, to be ferreted out and branded with red, white, and blue letters; honest and intelligent criticism often was stigmatized as disloyalty. Perspective was warped; we were myopic, dogmatic, intolerant, and authoritarian.

Authoritarianism jeopardizes scientific progress wherever and whenever it occurs—in the Union of Soviet Socialist Republics or in the United States of America, in the Dark Ages or in the Twentieth Century. It seems incredible that in our generation the law of organic evolution could be amended by decree in the Union of Soviet Socialist Republics and repealed by statute in two states of the United States of America, especially after this enlightening and useful scientific law had been centuries in evolving.

This was just a few years before the anticommunist witch-hunting by Senator McCarthy.

An article by Stakman in the *Scientific Monthly*[3] carried this statement at its head: "Dr. Stakman has recently returned from a scientific mission to Japan. A group composed of five American scientists reviewed, under the auspices of the National Academy of Sciences, progress being made by Japanese scientists." The article starts:

> Agriculture is fundamental to human subsistence, an obvious fact too often ignored by our lawmakers, educators, and the general public itself. A primary service that science must render to agriculture is to emphasize the importance and peculiar character of this basic industry, for indeed man is dependent on plant growth—on photosynthesis—for his very existence on earth. All his food and much of his raiment and shelter come either directly or indirectly from plants. Animals, important though they are, are essentially transformers of plant products and not primary producers.

This statement is found almost verbatim in a number of Stakman's talks and writings of this time. It is perfectly true, of course, and since he was talking to or writing for different groups, he had to repeat this to get it across. He continues:

> Realism and common sense are beginning to replace the visionary optimism of the boomers and boosters who had the naive concept that land in the United States was unlimited in quantity and imperishable in quality. Time was when land was so plentiful that it could be exploited without endangering the national welfare. But that time is long past.
>
> The pioneers who cut the forests, plowed the grasslands, drained the swamps, established the cattleman's empire, and discovered the sheepman's paradise aided in the phenomenal development of the United States. But they reckoned too little with the future and the penalties involved in the destruction of 50 million acres of cropland and the virtual destruction of an additional 50 million: 100 million acres of land—enough to feed and clothe 50 million people moderately well.

This makes it seem that Stakman may have believed that merely feeding and clothing another 50 million people is somehow a virtue—as if the ultimate good would be to feed and clothe moderately well as many millions or billions as our total exploitation of all possible resources of the world might enable us to do. Stakman did not subscribe to any such doctrine. He continued:

> We now have in the United States about 400 million acres of cropland, exclusive of grazing lands not suitable for cultivation—about enough to feed and clothe 160 million people with our present standard of living and productive efficiency. Much of this land is subject to erosion; common decency to the nation and its future generations demands that everything possible be done to protect and preserve this national resource, and that it be done promptly and scientifically.

Stakman was not alone in preaching the sermon of conservation of land; even then it was a chorus, if a somewhat muted one.

His talk on the importance of science was given several times, but the only published version I could find was in the British journal *Research*.[4] The general theme of the paper is that science is necessary, that we need more science, not less. Science, and especially some of its applications, has always come under criticism, and always will, from some people who, while surrounded by and dependent on things provided by science, decry the science that provided them. Few choose to return to the caves.

Not only is civilization always on trial, but the record is that many civilizations have failed; even the vaunted western civilization has several times been on the brink. Was it the fault of science? [The] record shows that it was not. Was science, for example, responsible for the Dark Ages? Truth is that lack of science rather than too much science was largely responsible. The authoritarianism that suppressed individual thought and stifled intellectual progress was as unscientific as dangerous. It could be dangerous even today.

Should there be a moratorium on science, as is sometimes facetiously or seriously proposed? The dire consequences of the virtual blackout of science in the Middle Ages are known to every casual student of history. We smile indulgently when we recall that learned professors in the best universities solemnly but harmlessly taught that the elephant had no joints in his legs, that the legs of a fox were shorter on one side than the other, that old eagles periodically visited the fountain of youth to become young again. We are inclined to wonder at the wide acceptance of the doctrine that the function of a learned mind was not to find out whether certain doctrines were true but to adduce arguments to prove that they were true. The blackout of science had many tragic consequences also. The abasement of the body to purify the soul, the reliance on miracle and fetish healing, and the consequent lack of sanitation, led to some of the most terrible disease epidemics in the history of mankind. Perhaps the most tragic of all is the fact that much of the misery and many of the fears resulted from retrogression—in surgery, in sanitation, in engineering, in a natural and rational attitude toward life and problems of living. If time blunts the acuteness of our realization of the inhumanity resulting from this historic moratorium on science we need only study the distressing situations in many scientifically backward countries today. Is the modern world entirely free from the blighting effects of authoritarianism and its tendency to suppress freedom of inquiry and expression?

What is needed is not a moratorium on science but even more science and technology; particularly is there need for man to try to become objective with respect to himself and the group of which he is a part. We need to emancipate ourselves from our own stupidity and ineptitude with respect to ourselves in social affairs. Even though the methods of social sciences may not be as precise as in some fields of natural science, is it too much to hope that man can learn to substitute facts and principles for unsupported assertions and easy generalizations? Can man learn to demand truth instead of designing propaganda? . . . The emancipation of man from his baser self obviously requires science; but not only science, it requires ethics also. . . .

Science can contribute as much to the refinement of the spirit as to the enlightenment of the mind. For is it not irrational and completely unscientific to ignore the facts and situations that make people and peoples act as they do? Would we always be so ready to condemn other people and peoples if we studied their problems with scientific objectivity and based our judgments on facts instead of prejudices? If we used as rigorous criteria of truth in human affairs as every good scientist uses in his science, would we tolerate the ethics of the trickster and the demagogue in human relations? If we always faced the truth, would we be as Pharisaical as we often are? Does not rigid insistence on truth help to develop a sense of justice and conduce to ethical conduct? Science helps to humanize but science needs help in the humanization process. There can be no moratorium on science.

As most of us know, it is much easier to be reasonable and rational and ethical when we are alone in our study in the evening than when we are coping with the people and problems we encounter during the day. This does not detract from Stakman's argument that science should make man more reasonable and rational.

In writing about international understanding,[5] Stakman contends:

> Science is a part of culture; it is one of the essential humanizing agencies. "It helps man to understand and modify his environment; it helps him to accomplish more with less effort; it helps him to develop wisdom; it contributes to rational and ethical concepts and conduct; it can contribute to international understanding and to co-operation among peoples; it can contribute to the attainment, maintenance, and enrichment of peace.
>
> "The obligations of science are even greater than its accomplishments. Science must contribute to society as well as to science; it must serve as well as enlighten." [Quoted from Sarton, "The Life of Science," U.S. Nat. Comm., UNESCO News, July 1951.]
>
> The objectives and values of science in education and in the progress of civilization are multiple and need not be mutually antagonistic. The spirit of free inquiry, of experimentation, of research, is essential to progress. Curiosity should be encouraged, not curbed. There are individuals with talent for discovery; their efforts should be supported, not regimented. Suppression of curiosity and prescription of aims and methods in scientific investigation are likely to block progress and lead to retrogression. The spirit of inquiry is important, but the spirit of service is important also. And the two need not be incompatible. Some scientists must use their knowledge and skills in improving technologies, in improving the living conditions of mankind, in contributing to the solution of urgent human problems.
>
> Surely one of the most urgent of all human problems is that of promoting understanding and co-operation among peoples. . . . it might be presumptuous of a natural scientist to discuss the reasons for tensions and contentions that lead to war. It sometimes is difficult to understand why reasonable compromise seems to be impossible even between presumably educated individuals, especially when compromise would be both reasonable and advantageous. Some individuals do not adjust easily to people and situations; some have genius for creating misunderstanding, and some have genius for misunderstanding. It sometimes is difficult to find even small areas of agreement among individuals. And the difficulty is likely to be still greater among peoples.
>
> Misunderstandings among peoples often result from ignorance rather than viciousness. Knowledge does not guarantee sympathy, but it is basic to it if sympathy connotes "reciprocal liking and understanding arising from

community of interests, community of aims and compatibility of temperaments." Peoples may co-operate successfully if their common aims and interests are strong enough, despite temperamental incompatibility. Do the peoples of the world have community of interests and aims that can be satisfied better by co-operation than by contention?

Among most people of the world the problems of subsistence and continued existence are primary and vital. "Give us this day our daily bread" has deep meaning to more than half the people of the world who hunger or even face the threat of starvation. The life expectancy of large groups of peoples is less than thirty years. Fear of hunger, fear of disease and premature death plague hundreds of millions of people in many countries. They are directly and vitally concerned with food, clothing, shelter, and health.

Why is international co-operation important? Among the greatest menaces to food production are bad weather, plant disease, insect pests, and diseases of domestic animals, none of which respect international boundaries. Stem rust of wheat can be taken as an example.

Stem rust of wheat is a devastatingly destructive plant disease that is international in character and therefore requires international co-operation for its control. Stem rust is caused by a fungus that multiplies by means of several kinds of spores; it reproduces asexually on certain cereals and many grasses and produces its sexual stage on certain species of barberry bushes. Stem rust of wheat is only one variety of the species *Puccinia graminis* that infects wheat, barley, and some grasses. . . . The wheat variety of stem rust comprises more than 250 known parasitic races that differ in their ability to attack varieties of wheat. A wheat may be immune from some rust races, moderately susceptible to others, and completely susceptible to still others. And new races are continually being produced by hybridization between existing races in the sexual stage of the rust on susceptible barberry bushes. Obviously, then, there are two principal methods of controlling stem rust: eradication of rust-susceptible species of barberry, and the breeding of rust resistant varieties of wheat.

Rust resistant varieties of wheat have been produced in the United States, Canada, Mexico, and other countries, but none have remained permanently resistant because the rust has always bred new races to attack them. It is known that a rust race that becomes established anywhere in North America is likely eventually to become established everywhere in North America, because about 70 billion spores may be produced on a single barberry bush, and these can be blown by the wind to wheat, where they can cause such abundant infection as to produce 50 thousand billion spores on a single acre. These spores are so small and light, about 1/1000 inch long, that they can be carried far and wide by the wind, making Mexico, the United States, and Canada one vast area in which the rust can spread and develop. The rust is an international traveler, and must be controlled through international cooperation. . . .

International co-operation then not only is desirable but in many cases it is essential. A case could be made for internationalism in most sciences. The need for international co-operation in helping solve problems of subsistence is of primary and vital concern to most peoples of the world. Therefore they should be able to understand the benefits derived from past co-operation on common problems, and the mutual benefits to be derived from common efforts.

Can recognition of the community of interest in problems of subsistence and health and of the values to be derived from co-operative effort lead to international understanding and good will? Surely it can help. But miracles need not be expected; evolution proceeds slowly. Cynics tend to magnify and visionaries to minimize the obstacles to international understanding. There is no easy way to remove them. The hope is in education, in the kind of education that is not restricted to acquisition of knowledge and skills but that also helps individuals evolve toward intellectual enlightenment and spiritual refinement. To this kind of education all of the sciences and all of the humanities must contribute.

Whether the schoolmasters and school administrators to whom he was speaking followed him through the excellent summary of the epidemiology of stem rust may be questioned. But they certainly must have gotten the idea that Stakman knew what he was talking about, that international cooperation was necessary for disease control, and that he firmly believed that international cooperation in many activities other than disease control was desirable and essential. He had given the same speach to the Central Association of Science and Mathematics Teachers in Cleveland, Ohio, shortly before, and it was later given to other groups. Stakman could preach the same sermon over and over again, without sounding trite or perfunctory; each time, it sounded fresh and had the force of right and reason and strong conviction. He believed in what he said and hoped that others would, too.

The following is from a talk given at the Jubilee convention of the Association of Applied Biologists, in England, on some of the developments in plant pathology between 1904 and 1954:[6]

> My first tremendous impression of the terrible devastation caused by plant disease was in the first week of August 1904, when I stooked [surveyed] about 200 acres of wheat almost completely ruined by stem rust. My most recent one was in the first week of August 1954, when I saw 2 million acres of durum almost completely ruined by stem rust and 10 million acres of bread wheat badly damaged by it.
>
> In 1904 much was known about fungus pathogens, but very little was known about the bacteria and viruses that cause plant disease. In 1904 scarcely a dozen bacterial diseases were known; now at least 170 different species are known to attack 150 genera of plants belonging to fifty families. During the past half century more than 200 plant viruses have been discovered and studied on virtually all kinds of economic plants. Not only has a vast amount of empirical information been accumulated, but deeper understanding has been attained and far greater precision in methods of investigation have been developed.
>
> Many concepts regarding fungus pathogens have been developed since 1904. It is true that there had been considerable experimentation, that emphasis was shifting from the descriptive and taxonomic phases to the dynamics of fungi and fungus diseases. Mistakes were made; the mycoplasm theory, adaptation, bridging hosts, certain ideas regarding predisposition and the nature of disease resistance have been discarded. But today there are organized bodies of knowledge regarding the genetics, the physiology, and the ecology of plant pathogens. Enough information has been obtained to develop new principles and to validate many hypotheses. Sound concepts have replaced many unsupported theories. We are beginning to understand plant pathogenic fungi.

There has been a great advance in knowledge regarding plant parasitic nematodes, and new ones are continually being found. Realization of the importance of this group of pathogens is somewhat tardy, but there are indications that they are beginning to receive the attention that they deserve.

The nature of many non-parasitic diseases has been elucidated. The role of industrial by-products and of micro-elements in the soil has been determined in many cases, and the way has been pointed to the investigation of other diseases that probably are due to similar causes. It was scarcely suspected in 1904 that a few parts per million of chemicals such as selenium could be deleterious to plants. Nor was it known that the addition of small amounts of boron, sulphur, manganese, copper, zinc and other chemicals could remedy many disease conditions that were completely baffling 50 years ago.

Losses in harvested products during transportation and storage have been greatly reduced in many countries. In 1904, as an example, rots of oranges during shipment from California to the eastern seaboard of the United States often destroyed 25–50% of the fruit. There were similar losses in other perishable products. Science and technology have greatly reduced these losses. There still are important storage problems of many products. As an example, wheat and other small grains are likely to be generally infested with moulds of the *Aspergillus* and *Penicillium* group which remain dormant at relatively low moisture content. A fractional percentage increase in moisture, however, may permit them to grow and cause "sick wheat." With the basic facts now available, control becomes an engineering problem in large grain silos.

Information about "sick wheat" being caused by species of *Aspergillus* was fresh off the research fire in his own department; Stakman grasped its significance long before many of the commercial grain merchandising firms were willing to recognize that molds might be involved.

This was part of a speech delivered at a centennial symposium at Michigan State University in May 1955:[7]

And we can learn much about the blackout of scientific attitudes during much of the Middle Ages. There was more retrogression than progression. Intellectual authoritarianism was so strong that it was taught as late as the fifteenth century that the legs of the fox were shorter on one side than the other; that the elephant had no joints in his legs and could not lie down; that bees carried gravel as ballast when in flight; that the lion drew a circle around himself with his tail when he lay down to sleep, that no animal dared enter the circle, but that the lion slept with his eyes open. Could it be that the lion doubted what the professor taught and kept his eyes open to see if it were true? There are reasons why there might have been reluctance to examine bees for their gravel ballast, but foxes were killed for sport, and the professor could have measured the legs with safety and profit.

We smile indulgently at these teachings of the funny professor of the past. But there was nothing funny about the consequences of the almost complete suppression of a naturalistic and rational attitude toward life and living; the funny teachings of the teachers were only one symptom of orthodoxy, traditionalism, and authoritarianism that not only blighted the intellect but also the lives of human beings. Filth and squalor were considered inevitable. The result was epidemics of diseases such as the Black Death that decimated populations and filled life with terror and dread; and the only remedy was to burn nonconformists by the thousands. It was not the suppression of what science there was, it was the inhuman result of the suppression that was so tragic. Bigotry is no less dangerous today than it was 500 years ago. The new view of man in his biological environment is or should be the view of a man to whom truth about the facts of his environment is sacred. He can only learn the truth about this environment by studying it thoroughly and honestly, and the truth should be sanctified by its human values. . . .

Darwin and Wallace established the principles of organic evolution, which some philosophers, poets, and scientists had sensed dimly and expressed vaguely from the days of Aristotle onward. Aristotle to Darwin— more than 2000 years during which some men had been searching for truth regarding natural relationships among plants and animals and the mode of origin of new kinds. Handicapped by the dogma of special creation and fixity of species, and often by rigid authoritarianism, basic concepts developed slowly and often were expressed cautiously. But finally Darwin's *Origin of Species* appeared in 1859. Darwin not only stated that evolution was fact but explained how it came about through variation, natural selection in the struggle for existence, and the survival of the fit. This revolutionized ideas regarding the plants and animals in man's biological environment. . . .

Mankind is again at the crossroads, looking for the right road. Ours is said to be the first generation that is seriously concerned about the welfare of its grandchildren. Although this may not be true literally, there is evidence of confusion in the present and concern for the future. This may indicate that man realizes that he needs more knowledge, foresight, and wisdom. Man has encountered many crossroads during the course of his evolution toward civilization. Sometimes he has chosen well, sometimes badly. Races of man have become extinct, civilizations have decayed and collapsed. But civilization as such has survived many crises. The trend has been generally upward, with many temporary fluctuations, and some racial and national casualties along the way. The course of civilization apparently is not predetermined, automatic, and guaranteed to reach Utopia. It depends on man's own efforts, on man's intelligence, his will, his ethics, his wisdom. He has needed them in the past; he needs them now. Possibly he needs them more urgently now than ever before. Whether the present crisis is more dangerous than those of the past may be debatable; but there can scarcely be any debate about the fact that there is a present crisis.

Man is menacing himself with two tremendous powers: the power of atomic energy and the power of human reproduction. Man's intellect has given him partial control of tremendous energy that can be used with terribly destructive effect or with humanly beneficent effect. Man's will must determine what he will do with this energy. And he should will to act as wisely as possible.

The power of human reproduction also is tremendous. The population of the world is exploding in our faces. It is now about 2.5 billion and is increasing at a rate of one percent or more a year: at least 25 million new mouths to feed each year. At this rate there is another Australia to feed every 4 months, another Canada every 7 months, another United States every 6 years. In forty years, at the present rate of increase, the world will have to feed at least 3.5 billion people, possibly more.

The rate of increase is greatest in the newer countries of the Western Hemisphere, but the absolute increase is greatest in the densely populated countries of Asia. Despite a high death rate, the population of India and Pakistan increased by 48 million in the decade 1931–1941. Despite the death of 100 million from starvation in China during the past century, estimates are that the population will reach 950 million by the year 2000, only forty-five years in the future. Is there enough land and water in the world to feed and clothe the products of man's prolificacy?

Stakman's figures on population increases were taken from official estimates and forecasts, and the estimates were low: instead of 3.5 billion people by the year 2000, there were more than 4 billion by 1980, with 1 billion in China. Many demographers and biologists, including Stakman, recognized the danger of overpopulation decades ago.

In an article on the need for international cooperation,[8] Stakman described the spectacular increase in Race 15B of stem rust in the United States in 1950:

> The year 1950 was the most spectacular in the known history of wheat stem rust in North America. An extraordinary combination and sequence of meteorological events enabled Race 15B to increase from very small beginnings to a percentage prevalence of 27 percent, in contrast to the previous year—when it was not found among the 669 uredial isolates from the United States nor among the Mexican and Canadian isolates. In 1949, Race 15B was isolated only once—from aecia on barberry in Virginia. In the spring of 1950, Race 15B was first detected in small quantities in northern Texas, and it increased in prevalence as it spread northward to Canada. Owing to an exceptionally late crop season and a late fall, it became epidemic on late wheat and wild grasses; air mass movements then carried enormous numbers of urediospores southward and southwestward into winter-wheat areas of southern United States and far into Mexico, where the uredial stage overwintered and furnished inoculum for a northward movement in 1951, when the percentage prevalence was 41 percent. Thus this race became an international traveler in an astonishingly short time, having extended its range over an area of about 3 million square miles in little more than a year.

This was a prelude to an appeal for more international plant disease nurseries:

> More international rust nurseries are needed, in which resistant varieties produced in one country can be tested for resistance in all countries in a geographic area, to determine the possible variability in resistance to rusts and other diseases under different meteorologic and edaphic conditions and to give preliminary indications of the existence of physiologic races that may attack them. This has long been standard practice within most countries, but it has become increasingly apparent that tests must be expanded to a regional basis.

In a talk before the United Nations Conference on the Application of Science and Technology,[9] Stakman again expressed the need for international cooperation:

> International cooperation can aid greatly in improving disease control by means of resistant varieties. Varieties may be highly resistant to all parasitic races of the pertinent pathogen in one country but susceptible to those in

another. If the pathogen is not distributed by wind or other agents of dissemination beyond the control of man, they can be excluded by quarantines or other phytosanitary regulations. Those which are wind-borne, however, are likely to spread sooner or later to all countries within homogeneous geographic areas. International cooperation in testing and breeding of varieties can therefore be mutually beneficial. Indeed it is essential to most rapid and permanent progress.

Disease resistant varieties have been of incalculable value in all countries where available scientific knowledge and technical skills were utilized in breeding programs. In the United States, flax wilt, cabbage yellows, and several other wilt diseases have been kept under control for almost 50 years; the sugar beet industry of western United States was saved from annihilation by means of varieties resistant to virus curly top; and losses from some major diseases of virtually all major crop plants have been substantially reduced. The spring wheat area of northern United States and Canada has been protected against the scourge of stem rust in about 20 years of the past 40, not a perfect record, but a good one. Within the past 20 years, wheat production has been doubled in Mexico as a result of the production of superior, rust-resistant varieties and better cultural methods.

Stakman, in 1964, writing about plant pathology,[10] said:

My point of view on plant pathology is that it is the most fascinating of all sciences to plant pathologists specifically and one of the most important of all sciences to the peoples of the world generally.... Why does society support plant pathology, and what is its obligation?

Quoting from myself, "For better or for worse, plant pathology had its genesis in fields and granaries more than in halls of ivy. Society needed agriculture and agriculture needed plant pathology. Of course, plant diseases excited curiosity, but the curiosity was more excited because the diseases struck in vital spots, the bread baskets and the fields that filled them. It was essential to social progress that the baskets be filled, and men of scientific bent essayed to fill them." This emphasis on the practical is nothing to be ashamed of; on the contrary, plant pathology can be proud that it is really needed and not merely tolerated by society. The need has existed throughout recorded history, and, although it was acute in some places and at some times in the past, it is doubtful whether the global need ever was greater than it is right now.

That the food problem is one of the gravest now confronting mankind is obvious to all who have observed present conditions or studied pertinent data about them. Man cannot run away from the problem, nor can men solve it by pious platitudes about "the dignity of man," "freedom from want," and the virtues of their cherished politico-economic systems. The problem is vitally real, and it can be solved only by means of realistic and competent action.

Stakman here was writing for a wider audience than just plant pathologists. No one knew better than he that plant pathology was only one of a dozen or more disciplines involved in increasing and maintaining stability in food production for an ever more rapidly increasing and hungry population in the developing countries. And he also knew that increased food production would lead only to increased misery in the future if it led only to an increase in numbers of people, who needed more food to produce still more people. He was writing for all concerned people, everywhere.

On May 13, 1964, in Washington, D.C., Stakman was given the first Cosmos Club Award (Figure 18). In his acceptance speech,[11] he said:

> Good sense should be the tie that everywhere binds science and society together in a common effort for human enlightenment and human betterment. For "Good sense is, of all things among men, the most equally distributed; for everyone thinks himself so abundantly provided with it, that those even who are most difficult to satisfy in everything else, do not usually desire a larger measure of this quality than they already possess." Thus wrote René Descartes more than 300 years ago in his "Discourse on the method of rightly conducting the reason, and seeking truth in the sciences."...
>
> It is sometimes said that modern science began with Descartes, who lived from 1596 to 1650. However that may be, the development of science was a long evolutionary process, with its faintest beginnings dating far back into prehistory. Why did it evolve?
>
> Science evolved because man is a curious and relatively intelligent animal. Primitive man had to be curious in order to survive; he had to use his senses in searching among many plants and animals to find those that were edible and nutritious and those that might sicken or kill him. Although he learned much about natural objects, there were many natural phenomena that he could not understand. What caused the sun to shine, the rain to fall, the thunder to roll, the lightning to strike? As he could neither understand nor control many awesome natural phenomena, he not unnaturally resorted to mystical explanations; he created supernatural beings in an enlarged image of himself. He created his numerous pagan gods and developed codes and rituals to keep them in good temper so that they would help rather than harm him and his. Naturally enough, some men cleverly convinced the populace that they had special influence with the gods; hence there emerged a guild of influence peddlers—the shamans, medicine men, the magicians, and the soothsayers—, a sort of primitive priestly caste. Many of these men undoubtedly were exceptionally observant and shrewd, as is true of medicine men among some primitive peoples today. They had learned certain scientific facts about man and nature but kept them secret in order to reinforce their pretensions to magical powers....
>
> To become intellectually enlightened man had to develop the desire to understand his world, not only to live better in it. He needed to substitute sense and reason for mysticism in elucidating the nature of natural phenomena. He began to approach the goal a scant two hundred years ago, although the Greeks made a good run for it more than two thousand years ago.

Stakman then takes up educational systems from the ancient Greeks through scholasticism and the Dark Ages, and continues with:

> An intellectual revival began in the twelfth and thirteenth centuries. The medieval universities began to develop: Salerno, starting with medicine; Bologna, with Roman law; and Paris, with theology. The medical school at Salerno was incorporated into the University of Naples, established by royal charter in 1224, "in order that those who hunger for knowledge may find

Citation

ELVIN CHARLES STAKMAN

DISTINGUISHED BIOLOGIST, GREAT TEACHER, AND STATESMAN OF SCIENCE

Scientist, linguist, and historian, he is recognized internationally for new knowledge brought to the biological sciences and for his inspired manner of guiding its application for the benefit of mankind.

His distinguished contributions to the field of plant pathology have been multiplied manyfold by the students who have come to him from every part of the world to profit from his knowledge and wisdom and who in turn have introduced new generations of students to the "Stakman philosophy."

Following a long and brilliant career as teacher, investigator, and administrator at the University of Minnesota, he is currently engaged in a second career—the advancement of the agricultural sciences on the international front.

As a statesman of science he has always been a lucid exponent of a humanistic scientific philosophy. His great intellectual and scientific merit has been recognized on many occasions by national and international societies, academies, and boards which have elected him to membership and high positions. Many honorary degrees and other awards from institutions in this country and abroad have been conferred upon him in acknowledgment of his outstanding achievements.

For his clear voice in the cause of science in the service of humanity, the Cosmos Club is proud to designate Elvin Charles Stakman as the first recipient of the Cosmos Club Award.

Figure 18. Citation given to Stakman by the Cosmos Club when he received the Cosmos Club Award, 1984. (Courtesy, Cosmos Club, Washington, DC)

within the kingdom the food for which they yearn, and may not be forced to go into exile to beg the bread of learning in strange lands." A worthy sentiment indeed! But what kind of knowledge did the hungry get? For the most part they got scholasticism—the method of the Schoolmen.

He continues with the history of education, with quotations from Comenius, Rousseau, Pestalozzi, and Paulus Vergerius. On occasions such as this when a modicum of showy scholarship was in good form, Stakman loved to show off a little; if he had wanted to really strew pearls, he probably could have quoted from all these men in the languages in which they wrote, verbatim, correctly, and from memory only slightly aided by his extensive card file kept for the purpose. He continues:

The Golden Age of Biology too began to yield rich and abundant fruits a hundred years ago. For the first time in all the millenia of his existence man began to understand bacteria and other microorganisms. He learned how many of them caused some of the most fearsome diseases of his food plants, his animals, and of himself. And he learned how to control many of the worst ones by sanitation and public health measures and by miracle drugs and antibiotics. He learned how to use many microorganisms in enriching the soil on his farms, in producing foods and flavors, in industrial processes, and in producing antibiotics to fight disease in ways almost undreamed of a few decades ago. Darwin's law of organic evolution, conceived in crude form 2000 years earlier, revolutionized man's outlook on nature; as nature was not fixed but in [a] constant state of change, it was natural to think that man could improve his plants, his animals, and even man himself. And he has been remarkably successful—at least with his plants and animals.

Man has even learned much about the nature of life itself. He has at least removed the dark fog of mysticism from many living phenomena, is rapidly clarifying his vision about many more, and may yet learn to understand and control his own behavior better. And yet he needs to learn much more, for there are many grave problems of human subsistence, human health, and human relations in his crowded and turbulent world.

Man's material future depends largely on his provision and control of energy, for power, for weapons, and for food. Science gave society atomic energy, and society either can use it wisely or misuse it unwisely. Science also gave society the means to produce food and protect health better than ever before. Will societies use these means fully but sensibly? Or will they multiply their numbers faster than they know how to feed them?...

... Some countries need good scientists; others need to give their scientists a chance to work effectively. Smothering bureaucracy and strangulating red tape are stifling scientific effort in many countries today. And, unfortunately, some agricultural scientists disdain to investigate anything so close to earth as agriculture. Do true scientists sully their science by using it to help hungry humans to get their daily bread? Many are the reasons why some hungry countries are hungry....

Food and health are two of the most elemental human wants, and international cooperation is essential in satisfying them as quickly and as widely as possible. Is it not simple humanity and good sense for all societies to cooperate in concentrated efforts to alleviate hunger and suffering of human beings? Perhaps collaboration on such vital human problems could conduce to mutual respect and to more amicable international relations.

Zeal for a common purpose can submerge many differences between the zealots.

Science has contributed much to the material welfare and to the intellectual enlightenment of society. Can it also contribute to spiritual refinement and the improvement of human relations? Can it help promote the general use of the scientific code of scrupulous intellectual integrity within each society and between all of them? Can it help to eradicate intolerance and bigotry from societies of the world as it has helped eradicate malaria, cholera, and yellow fever from many areas of the world? It can do its part, but its part is only part of what is needed. Society and all its civilizing agencies must do their part also, and may good sense, and even better sense, prevail in all their efforts!

Usually by the time a man is approaching four-score years his optimism is getting frayed at the edges and dirtied with blotches of frustration and despair; publicly, at least, Stakman, who saw as clearly as anyone and far more clearly than most the hazards and pitfalls that we face, maintained a hopefulness and even a belief that reason and decency might eventually prevail, that the promise of a good life might be realizable and realized.

In 1969, Stakman wrote on agricultural education:[12]

Eventually every country must adjust its population to its means of subsistence; it must produce enough food for its people or develop the purchasing power to buy it—provided enough food is available for purchase. But will enough always be available? Some thoughtful people say that it will not, unless the rate of population increase can be reduced rapidly and drastically or food production can be increased quickly and substantially.

If present food-population trends continue, the world food situation will be critical and may be catastrophic within two decades, according to many competent students of the problem. Among neo-Malthusians there appears to be a growing conviction that Malthus probably was more right than wrong....

Man has become the potential master of his own fate in his struggle for subsistence. Thanks to science and technology, he has the means to restrict his numbers simply and humanely and to increase his food supplies quickly and substantially. But to become the actual master of his fate, he must have the will and wisdom to utilize fully his present means and the wit to devise better ones for the future.

Concerning teaching:

Teachers tend to become too unimaginative, too apathetic, or too timid to be unorthodox in concepts or in methods; they tend to become conformists. It requires courage to be unorthodox, and it requires intellectual initiative and intelligence to become successfully unorthodox. But why change old ways for new ways unless the new ways are better than the old? We should learn from the experiences of the past and from the experiments of the present; we should be wise enough to be bold without being rash, for the need for effective education was never greater than now.

Even at almost 85, he was chiding teachers who become set in their ways and unimaginative, and was recommending that they develop new and unorthodox approaches to teaching, tempered with good judgment. Later he said:

It must seem obvious to most of those teachers who study students while trying to help them in their studies that good education has only limited power to overcome poor heredity. Or, in slightly more scientific terms, the genes for basic intelligence impose the limits within which education and other environmental factors can be effective. And yet in the United States of America in 1969 many environmentalists apparently are shocked by a recently published paper embodying the results of extensive studies that lead to the same conclusion. Some extreme environmentalists seem to think that the word "gene" should be expunged from the dictionary and that genetic laws should be declared unconstitutional, at least as applied to human beings. But this would open the way to more educational crimes. And there are too many now. . . .

It is a gross misdemeanor to create the impression that education is more potent than it really is. Education cannot do more than help individuals to develop to the limit of their inherent capacities. It is a disservice to the individual to overrate him. And it is a disservice to society to underrate the importance of heredity. In a democracy every individual should be encouraged and helped to develop his intelligence and skills to the limit of his potential, but it is futile and frustrating to try to force him beyond his potential. Education can do much, but it cannot give an individual a new set of genes. . . .

It is also an educational misdemeanor for scientists to quarrel about the relative merits and values of pure and applied science. Perhaps science itself needs some humanizing. To some pure scientists the degree of respectability in science is in inverse ratio to its practical human values. Presumably the "pure" scientist is motivated by curiosity, and curiosity should be kept alive; but people must be kept alive and healthy also, and science must do its part. Sometimes scientists, therefore, must also be motivated by a desire to render social service. Who shall say that one is a higher type of activity than the other? . . .

Whatever our basic concepts, the aim of higher education in general should be to develop a sense of scholarship. Higher education in agriculture has a dual obligation: (1) to promote productive scholarship—"the habitual application of an adequately informed and self-disciplined mind to the solution of problems"—especially when supported by a high standard of ethics and given moral purpose by a desire to render social service; (2) to capacitate technologists, teachers, and extension specialists to transmit the results of research to farmers and others in the most effective way.

CHAPTER FOURTEEN

Personal Glimpses

For six months in 1940–1941, Stakman was leader of a rubber survey party sponsored by the United States Department of Agriculture that explored for native rubber plants in northwestern South America. The oral history describes how he made the best of some uncomfortable situations.

I saw plenty of hunger in Latin America. I had started going to Mexico in 1917, and if a person got off of the main highway outside of the cities and off of the main tourist routes, he saw plenty of hunger. . . . so in 1940 in the rubber trip to some of the South American countries, Panama, Colombia, Ecuador, Peru, part of Brazil, you saw plenty of hunger, altogether too much of it. As a matter of fact, I was hungry there myself, and so I know something about what hunger meant. I was hungry for several reasons. In certain areas their food was not fit to eat, at other times it just wasn't available. My own hunger helped me to learn something about Spanish grammar, because we were out in the Choco, so-called. That was a federal territory in Colombia. It was very unsettled, some of the wildest part of the country on the tributaries of the Rio Atrata, and it was quite primitive country.

Well, this was on a rubber expedition party, and we spent Christmas Eve in a jail because it was the only place to stay in a little village comprised entirely of Negroes and Indians, on a beach, and we had to wait there for a boat that was going to pick us up to take us to another place where we caught a coast-wise boat back to Buena Ventura, and so we had some rice for supper that night. There was no hotel, of course, or anything of that type and the jail was the only place to stay.

The next day was Christmas, and we spent Christmas on the beach in a bamboo hut that had just been vacated by some primitive Indians, and the only thing that we had to eat was rice and coconut milk. We'd picked up a couple of coconuts on the way to this place on Christmas day, and everybody told us that there was no hurry because they said that everybody on the boat would have been celebrating Christmas and the boat surely wouldn't get there until the next day. It turned out to be that way. But in the meantime, we were hungry, because just plain boiled rice and coconut milk wasn't exactly what one would like for a Christmas dinner.

And added to that there was the aggravation that we had hammocks, because it was much safer to swing a hammock from a tree trunk and sleep in that than it was to sleep on the ground, where there were a lot of venomous reptiles. At least we thought so. [Stakman had a powerful fear of venomous reptiles, even in regions such as northern Minnesota where they do not occur; pushing through the Amazon jungle, where venomous reptiles do

127

occur, must at times have been a horror for him.] And so we swung our hammocks in this hut that had just been vacated by the Indians, and there were pigs that were rubbing their backs on the floor boards, because it was built up a little above the ground, and the fleas came up and the fleas bit us. Well, we had visions of really shanghaiing one of those pigs and getting some Christmas bacon or something, but there were too many people around so we didn't.

But I said that my hunger helped me to learn Spanish grammar. When a person is out in the tropics, and if you're out in a dugout canoe and are living only on plantains—that is, cooking bananas—that are cooked in river water and maybe have a little rice and you're lucky if you have a little salt, you get pretty hungry. Then if you yourself are helping to paddle the canoe, and the temperature is up in the 90s and the humidity about 90%, and you perspire very freely, you'd give a month's salary for a freezer full of ice cream. A Colombian agrónomo [agonomist] was along on the trip with us and he wanted to learn English and I wanted to polish up on Spanish, so I learned to use the past subjunctive in a condition contrary to fact, followed by the conditional mood in the conclusion. It means that the mood was quite conditional at certain times and also that the mood could be quite contrary at times. What I really said to him was (you could vary it in various ways) "If I had an ice house full of a ton of ice cream, I'd like to give you a couple of ounces." Or you could also say, "If I had a brewery full of ice cold beer, I'd like to give you a couple of droplets." Then of course he'd reply in kind.

I doubt whether many people, under those conditions, would have been interested in refining their grasp of the subtleties of the subjunctive mood in Spanish verbs. That Stakman was, is typical of him. It probably can be taken for granted that he mastered the verbs.

Athletics

Stakman was blessed with a body about as tough and agile as his mind, and he combined this with a strong competitive instinct that bordered on the belligerent. He was an enthusiast of just about all sports and athletic endeavors, but he especially liked the contact sports with plenty of physical give and take. So far as is known, he himself did not engage in any of these sports except handball, but in his informal discussions he certainly led listeners to believe that he was a boxer of no mean ability and was more or less intimate with both Mike and Tommy Gibbons (high-ranking pros in St. Paul) and that he was quite a basketball player. In the oral history, he says his arms were too short for him to have been a big league baseball pitcher, the implication being that all he needed was a little longer reach and he would have mowed them down. He played an effective game of handball, but his play, if not outright dirty, certainly was intimidating. If he was unable to reach the ball for a return, he would either accuse his opponent of blocking or else crash him into the wall, claim a foul, and have the point played over. He himself blocked whenever he could do so, then glared at his opponent as if daring him to call it. How good he would have been playing someone not in his own department and not under his thumb, with an impartial referee, is perhaps open to question.

He was a keen student of football and for many years was an enthusiastic follower of the Golden Gophers (the University of Minnesota team) and later of the pros. Not many football players were found in the College of Agriculture, but for a time in the 1920s, the School of Forestry, in the College of Agriculture, Forestry and Home Economics, was an academic refuge for varsity football players. If, during their football careers, they progressed through the academic ranks to the status of sophomores, as some of them did, one course they were required to take was forest pathology. Although Stakman was not nominally in charge of the instruction in forest pathology, as head of the department he arrogated to himself the right to decide on the grades for these gridiron thunderbolts, probably in violation of university rules and regulations. They all passed.

In the Stakman Day *Aurora*, I described Stakman's softball teams:[1]

The softball teams of Plant Pathology from the mid-1920s to the late 1930s were famous. Through those years Dr. Stakman was owner, general manager, chief scout and coach of these teams. Softball was not taken lightly in the Tottering Tower in those days. As soon as the athletic field became playable in the spring, usually around mid-April, practices were instigated, and these probably were every bit as well organized as the big-league baseball camps of the day, maybe even better. The first-string members, who had won their places in the previous years, worked at their specialties. The rookies were tried out at various positions, to see where they would best fit in. The practices ordinarily ran from 12:30 to 3:00 or 3:30 p.m., and often continued in the halls of the Tottering Tower until a fairly late hour in the evening. Classes were scheduled so as to avoid conflict with softball practice. As coach, Dr. Stakman either pitched or batted, in practice, and in either case he called the balls and strikes; if, when batting, he failed to hit a pitch solidly, it was a ball. As pitcher he had a mean, high-velocity sidearm delivery that was patently illegal, but which he stoutly maintained was totally legitimate. Anyone who could hit what he threw was likely to have a pretty good batting average against the ordinary pitchers in the league. Plant Pathology usually won the championship of the St. Paul Campus Interdepartmental League. They should have, because the team was solidly grounded in all the fundamentals of the game; they could field and hit, and when an opposing runner was on base, they knew which base to throw the ball to and how to throw it. As they played it, softball was not a harum-scarum game.

Dr. Stakman and Dr. Hayes, then head of Agronomy and Plant Genetics, carried their friendly rivalry to the softball field. One year Dr. Stakman hired, for part-time work in the greenhouse (thus making him eligible for the Department team), a pitcher from one of the Public Park Leagues in St. Paul—a man of muscle and sinew about 6 feet six and of somewhat surly disposition, which showed when he pitched. He intimidated the more gentle among the academics. He threw a cannonball; players on opposing teams trembled as they came to bat, as well they might, and few of them got so much as a single off this ringer. Dr. Hayes the next year countered by bringing in a ringer of his own, from the Southwest Experiment Station, also a pitcher, by which time our semipro had gone to greener softball fields. Plant Pathology won the championship again that year, showing, according to Dr. Stakman, that Righteousness and Virtue will triumph over Villainy and Deceit, especially if aided by ability, long practice, determination, and a modicum of guile. The idea was to win. The old saw of Grantland Rice to the effect that "It isn't whether you won or lost, but how you played the game" was put in its proper perspective as idealistic mush and twaddle, perhaps appropriate in the Ivy League, but not to be taken seriously in real life. If, while winning, some of the players developed some character, fine, but certainly no one who played for or with Dr. Stakman on or off the field ever got the idea that losing was pleasurable or acceptable.

Stakman and Louise

The experiment station bulletin listed E. Louise Jensen as mycologist on the staff of the Division of Plant Pathology in 1913 and subsequently, when staff members were listed, up until the time of her marriage to Stakman in the fall of 1917.

Stakman must have begun to court her almost as soon as she joined the staff, since as early as 1914 he was writing her some fairly torrid notes whenever he was out of town, most of them of the my-heart-is-longing-for-you genre affected by swains of the day, and some of them in baby-talk, evidence of the pitiable condition to which strong men sometimes are reduced by love. Some of these letters were addressed to her at the Jensen Lumber Co., in South Minneapolis; her father evidently was in the lumber business. The family must have been at least moderately well-to-do, since one letter to her from Stakman is to a Los Angeles address, where she was visiting. In those days, that was far travel. The Jensens also had a summer home on Lake Minnetonka. In the Stakman Day *Aurora*, Esther (Mrs. Arne G.) Tolaas says of Stakman's courting:[2]

> He had been spending some of his week ends at the Jensen summer home at Lake Minnetonka with Louise. They never left the department together and when they met in the office on Monday morning Stak would greet Louise as though he had not seen her during the week end and they had not just parted.

Stakman and Louise Jensen were coauthors of one paper, on infection experiments with timothy rust. She must have also helped him in teaching, since in a letter to her in the fall of 1914 he gives suggestions about material to be presented to the class in wood structure that he was teaching to the forestry students and also about materials for the laboratory in a plant pathology class.

In some of his letters to her he called her "Stella," but she was always known as "Louise." His pet names for her were "Juno" (because of her Junoesque figure) and "The Viking." She was slightly taller than he, a honey blonde, soft spoken, gracious, attractive, and feminine, a charming and easy hostess. At times, when one or two friends were present at their home with the two of them, she could hardly make a statement or express an opinion that Stakman did not immediately contradict, sometimes rather brusquely and with deprecatory and ridiculing comments regarding her competence to form an opinion on the business at hand. This practice was embarrassing to the guests, but if it bothered her, she gave no hint of annoyance. The reason for this deliberate rudeness to a woman he loved and respected remains hidden.

In the early days after their marriage, they lived in an apartment on Raymond Avenue, near the St. Paul campus. Stakman sometimes had the Thursday evening seminar there. In 1925, when Stakman was away on a long around-the-world trip to Australia, Southeast Asia, India, Egypt, and portions of Europe, she had their house built at 1411 Hythe Street (Figure 19), about a quarter of a mile from the campus and within easy walking distance of the Plant Pathology Building.

There they occasionally entertained small groups of friends, including some of the more congenial staff members and now and then a visiting plant pathologist of note. These were sociable get-togethers of people who knew and were at ease with one another, not stuffy and formal affairs; sometimes Stakman put on a chef's cap and apron and carved for a buffet dinner (Figure 20). After prohibition ended, they served alcoholic drinks, but Stakman was not much of a drinker and did not need the stimulus of alcohol in order to talk. One drink could last him most of the evening, although in his last few years he recognized the effectiveness of the predinner dry martini as an antidote to some of the tribulations of the day. He professed to be and probably was somewhat of an expert on white wines, but he had no prejudice against bourbon or Scotch if his guests so preferred. I doubt that he ever in his life consumed enough liquor at one time to put him out of full command of all his faculties. He was a genial, congenial, and hospitable host, and he monopolized the conversation completely, but it was almost always an interesting, thoughtful, and informative monologue.

Stakman and Louise were deeply attached to one another. She accompanied him on only a few of his many journeys and junkets. She went with him to the University of Halle in 1930–1931, when he taught there for a semester. Whether she enjoyed it greatly there may be questioned; Halle-an-der-Saale in the winter in a poorly heated apartment was something to be endured, not enjoyed. She later went along on one or two trips to Mexico and one to India, but mostly she preferred to hold the fort at home. It was an attractive home and she had friends and things she enjoyed doing; she did not feel abandoned or deserted.

Mrs. Tolaas, in the Stakman Day *Aurora*,[3] says,

Figure 19. The house that the Stakmans built at 1411 Hythe St. in St. Paul.

Stak used to say that he had a prenuptial agreement with Louise to the effect that he was to have no domestic responsibilities so that he could be free to pursue his scientific career. If there was such, Louise certainly lived up to her part of it. However, he was a great kidder and a very charming one.

Louise died of a sudden heart attack at home one evening in 1962 with no advance warning. Stakman was there at the time. It was a severe blow to him. For some time after that when he came into our laboratory at noon to join Carl Eide and me at our frugal lunch, he would sit down, gaze out the window silently, and a couple of tears would course down his cheeks. He would wipe them away without embarrassment or comment. He was with us there in 1974 at noon on the anniversary of her death, and he said, "Louise died twelve years ago today." He continued to live in the home on Hythe Street, alone, and at times it must have been lonely beyond words, with the house full of memories of their 45 good years together.

Retirement

In September 1953, shortly after Stakman's official retirement as Head of the Department of Plant Pathology, a group of former students, associates, and admirers established the E. C. Stakman Award, in commemoration of his name and his contributions to plant pathology and to learning. The award, which consists of a scroll, a medal, and an honorarium, is given periodically to an individual who, in the opinion of the awards committee, has contributed significantly to plant pathology (Figure 21).

Figure 20. Christmas 1947. Left to right: Mrs. J. J. Christensen, Mrs. Don Fletcher, Stakman, Dr. Helen Hart.

After his retirement, Stakman retained and regularly occupied an office in the Plant Pathology Building (Room 305), which had been occupied by J. J. Christensen before the latter replaced Stakman as head. Stakman and Christensen remained as close after Stakman's retirment as they had been for many years before (during the previous 12 years, J. J. Christensen had always served as acting head during Stakman's many and prolonged absences), and the administration of the department continued to be, to some extent, a joint one. Stakman attended seminars regularly but sat in the back of the room and did not contribute as much to the tone of the seminars as before.

He continued as a consultant to the Rockefeller Foundation, making frequent trips to New York and taking occasional tours of duty in Mexico and shorter trips to other countries. Up until the early 1960s, he would often be in Mexico for months at a time, concerned with the international research programs and also with the promotion and development of the graduate school at Chapingo.

He was in demand as a speaker and traveled widely to interpret science to fellow scientists and to the public. Much of his writing during this period consisted of written versions of these speeches, portions of which are quoted in Chapter 13. He also now found time to summarize some of his knowledge and outlook in two books: *Principles of Plant Pathology*, with J. G. Harrar as coauthor, was published in 1957, and *Campaigns against Hunger*, written with his long-time friends and associates, R. Bradfield and P. C. Mangelsdorf, was published in 1967.

Figure 21. Stakman presenting the E. C. Stakman Award to José Vallega in 1960. Harold Macy, Dean of the Institute of Agriculture, Forestry and Home Economics at the University of Minnesota, is at right.

Last Years

Stakman kept the pleasant corner office in the Plant Pathology Building. The walls were lined with shelves holding bound copies of the many scientific journals to which he continued to subscribe, and the desk and table were piled high with books, correspondence, and reprints. He usually came to the office about 10:00 a.m. and soon went to the seminar room for coffee with the staff and graduate students. He stayed there, talking, until 12:00 or 12:30, then joined Dr. Eide and me in our laboratory for more talk and a single cup of tea, which sat there as he smoked his pipe and held forth on whatever topic occupied his mind that day. When we were ready to resume work, Stakman drank his tea, now stone cold, in a couple of gulps and went back to his office, where he read and wrote. He continued to attend Tuesday afternoon seminars.

From 1965 on, his traveling was greatly reduced, although he continued as a consultant to the Rockefeller Foundation. In 1970, he dictated his oral history, in New York. In 1973, he went to Chapingo, Mexico. Dr. M. de Lourdes de Bauer says in the Stakman Day *Aurora:*[4]

> On January 30, 1973, the Plant Pathology Department of the Colegio de Postgraduados at Chapingo dedicated to him its reading room. At the time he said, after hearing the presentation of some MS and PhD candidates, that he was pleased to learn the kind of research carried on at the Department, and appreciated the serious work done by the students with enthusiasm and ability.

In 1977, when he was 92, he traveled again to Mexico and was honored by the Instituto Nacional de Investigaciones Agricolas and by the Centro Internacional de Mejoramiento de Mais y Trigo at Ciudad Obregon and at the graduate school at Chapingo.

On July 19, 1977, a typically hot, humid, and uncomfortable midsummer day in Minnesota, alone at home, he suffered a paralytic stroke and congestive cardiac arrest. He was hospitalized, and a pacemaker was installed. This improved the heart function, but his physical activity was severely limited from then on. He could not walk without assistance and his speech was severely impaired. In late August, he was moved to his own home, with continuous nursing care. In March 1978, he was moved to a modern health care facility, the Presbyterian Home, on Lake Johanna, just north of St. Paul, where he had a large private room and expert nursing and medical care. Graduate students and other friends visited him frequently and he seemed to enjoy this.

On his 93rd birthday, a dinner party at a nearby restaurant was organized for him by the graduate students, and he attended in a wheel chair. The atmosphere was one of good fellowship and cheer, with nothing lugubrious about it, and Stakman appeared to have a wonderful time.

He continued in about the same condition until late December 1978, when he caught cold; pneumonia developed, and he was removed to a hospital. On January 21, 1979, a visitor found him seated on a chair watching a pro golf match on TV. Later on that day, he suffered a massive stroke, and he died at 11:25 a.m. on January 22. As specified in his will, his remains were cremated and the ashes interred, on January 26, in Sunset Memorial Park, near the graves of his mother and Mrs. Stakman.

His extensive library was left to be disposed of by the Department of Plant

Pathology, and his household goods were left to a sister-in-law. The residue of his estate, aside from some minor gifts to special friends, was left to the Department of Plant Pathology of the University of Minnesota. Including the proceeds from the sale of his home, this amounted to approximately $500,000. It was added to a permanent endowment fund established in the department, the income of which is to be used for such things as a memorial library, occasional visiting professors, and workshops and similar worthy causes for which public monies are not available. Even in death, Stakman contributed to the advancement of plant pathology and to the university that he loved and that he had served so long and so well.

HONORS AND AWARDS

Dr. Stakman held regular or honorary memberships in 13 scientific and learned societies in the United States and in Canada, Great Britain, Germany, India, and Japan. He was president of the American Phytopathological Society in 1922 and of the American Association for the Advancement of Science in 1949. In 1948, he was a member of the Scientific Mission to Japan under the auspices of the Supreme Command for Allied Powers. He was a member of the National Commission of UNESCO, 1950–1956; the Executive Committee, National Science Board, 1951–1954; and the Advisory Committee on Biology and Medicine, United States Atomic Energy Commission, 1948–1954 (chairman, 1953–1954, and consultant, 1954–1959). He held various offices in the National Research Council in 1931–1934, 1937–1938, 1947–1948, and 1950–1958.

He received honorary degrees from the University of Halle-Wittenberg, Halle-an-der-Saale, Germany, 1938; Yale University, 1950; the University of Rhode Island, 1953; the University of Minnesota, 1954; the University of Wisconsin, 1954; and Cambridge University, England, 1964.

Among the special honors given him were the Emil Christian Hansen Gold Medal and Prize, 1928 (for his contributions to the knowledge of physiologic specialization in fungi); the Medalla de Merite Agronomico, Colombia, South America, 1955; the Centennial Award, Michigan State College, 1955; the Certificate of Merit, Botanical Society of America, 1956; the Otto Appel Medal, 1957; the first Cosmos Club Award, 1964; and La Cruz de Boyacá, Colombia, South America, 1966.

A RECORD OF GAINFUL EMPLOYMENT

Job Title	Name and Location of Employer	Dates of Employment	Duties
Teacher	Red Wing High School, Red Wing, MN	1906–1907	Taught history, political science, German; coached baseball
Teacher	Mankato High School, Mankato, MN	1907–1908	Taught history, political science, botany, algebra; coached football, track
Superintendent of Schools	Argyle School Board, Argyle, MN	1908–1909	Teaching; supervision; coaching
Instructor in plant pathology	University of Minnesota, St. Paul	1909–1913	Teaching plant pathology and botany
Professor Chief of Division	University of Minnesota, St. Paul	1913–1940 ⎫ ⎬ 1940–1952 ⎭	Teaching plant pathology; research; administration of Section of Plant Pathology, 1913–1940
Collaborator (part time)	USDA, St. Paul, MN	1914–1918	Investigating physiologic specialization of cereal rusts
Pathologist (full time)	USDA, St. Paul, MN	1918	In charge of campaign for barberry eradication
		1919	Resumed work with University
Agent (part time)	USDA, St. Paul, MN	1917–1952	In charge of rust epidemiology studies
Scientific adviser	Firestone Plantations Co., Liberia	1930 (6 weeks)	Studied plantation rubber, made recommendations regarding research
Guest professor	University of Halle, Halle-an-der-Saale, Germany	1931 (1 semester)	Lectured in plant pathology
Leader, rubber survey party	USDA, northwestern South America	1940–1941 (6 months)	Surveyed area for native rubber plants; made government agreements

Job Title	Name and Location of Employer	Dates of Employment	Duties
Member, agricultural survey party	Rockefeller Foundation, Mexico	1941 (2 months)	Surveyed for agricultural needs; made recommendations for improvement program

(On basis of the above, a project was organized in Mexico in 1943, and Stakman was involved for one or two periods of employment each year thereafter.)

Member, agricultural and research survey party	Rockefeller Foundation, Central and South America	1947 (2 months)	Studied agricultural conditions and status of educational and research work in the natural sciences

Notes

Chapter 1

1. Stakman left his personal papers to the Department of Plant Pathology of the University of Minnesota. They include copies of many letters and speeches written throughout his lifetime. This letter and others in this book are from those papers and are quoted with the permission of Dr. David French, head of the department.

2. Throughout this book, quotations from Stakman that are otherwise unidentified are taken from his oral history, a work of 1,687 pages dictated in 1970 at the request of the Rockefeller Foundation and now kept in the Rockefeller Archive Center. Interviews for the oral history were conducted by a staff member of the Oral History Research Office of Columbia University. Quotations are used with permission of the Rockefeller Foundation.

3. It was then the School of Forestry, in the College of Agriculture, Forestry and Home Economics of the University of Minnesota.

4. Mrs. Tolaas's recollection was printed in the E. C. Stakman Day issue (pp. 11–12) of the *Aurora Sporealis*, the alumni publication of the Department of Plant Pathology of the University of Minnesota. This issue, published on May 17, 1979, on what would have been Stakman's 94th birthday, was devoted entirely to quotations from and reminiscences about Stakman, who had died on January 22, 1979.

Chapter 2

1. Dr. James later became the first dean of the College of Education at the University of Minnesota.

Chapter 3

1. A copy of this letter was among Stakman's private papers.

Chapter 4

1. E. M. Freeman, *Minnesota Plant Diseases*, The Pioneer Press, St. Paul, MN, 1905.

Chapter 5

1. This was Extension Bulletin No. 14 of the Minnesota Farmers' Library, published by the Department of Agriculture at the University of Minnesota. The authors were listed as E. M. Freeman, Ph.D., Professor, Vegetable Pathology and Botany, and E. C. Stakman, M.A., Assistant in Plant Pathology.

2. This was in March 1911. The authors were listed as A. G. Ruggles and E. C. Stackman—his name thus incorrectly spelled.

3. Stakman was listed as Assistant Plant Pathologist, and his name was spelled correctly.

4. Stakman was the sole author.

5. "Relation Between *Puccinia graminis* and Plants Highly Resistant to Its Attack," 1915, p. 198. Complete references for Stakman's writings are given in the section on his publications at the end of this book.

6. The quotation is from E. M. Freeman and E. C. Johnson, *The Rusts of Grains in the United States*, U.S. Department of Agriculture Plant Industry Bulletin 216, 1911. Stakman used it in a 1918 article (see note 8 below, p. 222).

7. "Infection Experiments with Timothy Rust," 1915, p. 215.

8. "Plasticity of Biologic Forms of *Puccinia graminis*," 1918, pp. 246 and 247; the second quotation is from p. 230.

9. "Can Biologic Forms of Stemrust on Wheat Change Rapidly Enough to Interfere with Breeding for Rust Resistance?" 1918, p. 122.

10. "Infection of Timothy by *Puccinia graminis*," 1916.

11. "Biologic Forms of *Puccinia graminis* on Cereals and Grasses," 1917.

12. Ibid., p. 492.

13. "The Occurrence of *Puccinia graminis tritici-compacti* in the Southern United States," 1918, p. 149.

14. There were two railroads running west out of St. Paul, both from the same depot, so he didn't have to roam very much.

15. "A Third Biologic Form of *Puccinia graminis* on Wheat," 1918.

16. "New Biologic Forms of *Puccinia graminis*," 1919.

17. "Wheat Stem Rust from the Standpoint of Plant Breeding," 1921.

18. *The Determination of Biologic Forms of* Puccinia graminis *on* Triticum *spp.*, 1922.

Chapter 6

1. See note 11 of chapter 5. The quotation, on p. 489 of that article, is from F. J. Pritchard, "A Preliminary Report on the Yearly Origin and Dissemination of *Puccinia graminis*," Botanical Gazette 52(3):169-192, 1911.

2. Stakman was thinking of Mexico and Central America, particularly, as possible sources of rust.

3. This was and still is the upper crust businessmen's club there.

4. "The Black Stem Rust and the Barberry," 1918. Reprinted separately as Separate No. 796, 1919.

5. Published by the Spring Wheat Improvement Association, Minneapolis, MN, 1920.

6. *Destroy the Common Barberry*, 1919. This bulletin was published in May 1919, and

revised editions were published in August 1919 and February 1923.

7. Published in 1923.

Chapter 7
1. See note 4 of chapter 6; the quotation is from p. 11.

2. "Spores in the Upper Air," 1923, pp. 599 and 604.

3. "Progress and Problems in Plant Pathology," 1955, p. 29.

Chapter 8
1. From Stakman's personal papers.

2. Hans Kniep, *Die Sexualität der niederen Pflanzen*. G. Fischer, Jena, Germany. 1928.

3. This involved inoculating corn plants with mixtures of the mated lines and then isolating from the resulting galls on the plants.

4. "Mutation and Hybridization in *Ustilago zeae*," 1929.

Chapter 9
1. Page 5.

2. Page 6.

3. Dr. F. H. Kaufert, who at that time was a graduate student in plant pathology, went along with Stakman and was his laboratory assistant at the University of Halle. He was also the first of several exchange students between the University of Halle and the University of Minnesota. Kaufert said that Stakman had trouble stopping when the two hours were up. That sounds reasonable.

4. Page 26.

5. Page 21. Originally from "Education: Needs and Virtues; Crimes and Misdemeanors." Pages ix–xxi in: *New Concepts in Agricultural Education in India*. A. S. Atwal, ed. Punjab Agricultural University Press, Ludhiana, Punjab, India. 1969.

Chapter 11
1. Letter to C. M. Christensen, August 11, 1981.

2. R. B. Fosdick, *The Rockefeller Foundation*. Harper & Row, New York, 1952. 336 pp. The quotation is from pp. 181–182.

3. "Agricultural Conditions and Problems in Mexico. Report of the Survey Commission of the Rockefeller Foundation." Page 1; second quotation, pp. 32–33. Typewritten report, 1941, used by permission of the Rockefeller Foundation.

4. W. W. Folwell, *History of Minnesota*, 2nd ed., Vol. 4, p. 87. Minnesota Historical Society, St. Paul, MN, 1930; 2nd ed., 1969.

5. E. C. Stakman, Richard Bradfield, and Paul C. Mangelsdorf, *Campaigns against*

Hunger. Copyright Belknap Press of Harvard University Press, Cambridge, MA, 1967. 328 pp. Reprinted by permission. This quotation is from p. 33.

6. Page 185.

7. From "Plant Pathology in Mexico," pp. 52–53. Copyright 1945 by Chronica Botanica. Used by permission of John Wiley & Sons, Inc.

8. Page 200.

Chapter 12
1. A 50-page report put out by the CGIAR in 1980.

Chapter 13
1. "International Problems in Plant Disease Control," 1947.

2. "Science and National Authoritarianism," 1950.

3. From "Science in the Service of Agriculture," 1949. Copyright 1949 by the American Association for the Advancement of Science; used by permission.

4. "Science—Its Sphere of Influence," 1950.

5. "Contributions of Science to International Understanding," 1952.

6. "Progress and Problems in Plant Pathology," 1955.

7. "The New View of Man in His Biological Environment," 1957.

8. "The Need for International Cooperation on Cereal Rusts," 1959.

9. "Contributions of Science to the Understanding and Control of Plant Diseases," 1963.

10. From "Opportunity and Obligation in Plant Pathology," 1964. Reproduced, with permission, from the *Annual Review of Phytopathology*, Vol. 2. Copyright 1964 by Annual Reviews, Inc.

11. "Science, Sense, and Society," 1964.

12. "Education: Needs and Virtues; Crimes and Misdemeanors." Pages ix–xxi in: *New Concepts in Agricultural Education in India.* A. S. Atwal, ed. Punjab Agricultural University Press, Ludhiana, Punjab, India. 1969.

Chapter 14
1. Pages 10–11.

2. Page 12.

3. Ibid.

4. Page 26.

PUBLICATIONS OF E. C. STAKMAN
1911–1966

The Smuts of Grain Crops. (with E. M. Freeman) Minn. Agric. Ext. Bull. 14. 1911.

Orchard and Garden Spraying. (with A. G. Ruggles) Minn. Agric. Ext. Stn. Bull. 121, 2d ed. 1911.

The Smuts of Grain Crops. (with E. M. Freeman) Minn. Agric. Exp. Stn. Bull. 122. 1911.

Potato Diseases. (with A. G. Tolaas) Minn. Agric. Ext. Bull. 35. 1912.

Spore Germinations of Cereal Smuts. Minn. Agric. Exp. Stn. Bull. 133. 1913.

The Smuts of Grain Crops. (with E. M. Freeman) Minn. Agric. Exp. Stn. Bull. 122 (rev.). 1914.

A preliminary report on the relation of grass rusts to the cereal rust problem. (Abstr.) Phytopathology 4:411. 1914.

The relation between *Puccinia graminis* and host plants immune to its attack. (Abstr.) Phytopathology 4:400. 1914.

A Study in Cereal Rusts: Physiological Races. Minn. Agric. Exp. Stn. Bull. 138. 1914.

The seed-potato plot. (with Richard Wellington) Minn. Farmer's Library, Vol. 5, No. 2. 1914.

A fruit spot of the wealthy apple. (with R. C. Rose) Phytopathology 4:333-336. 1914.

Spraying Calendar. (with A. G. Ruggles) Minn. Agric. Ext. Div. Spec. Bull. 1. 1915.

Experiments in the control of brown rot of plums. (with R. C. Rose) Minn. Hortic. 43:231-233. 1915.

Relation between *Puccinia graminis* and plants highly resistant to its attack. J. Agric. Res. 4:193-199. 1915.

Infection experiments with timothy rust. (with L. Jensen) J. Agric. Res. 5:211-216. 1915.

Spraying plums for brown rot. Minn. Hortic. 44:148-155. 1916.

Infection of timothy by *Puccinia graminis*. (with F. J. Piemeisel) J. Agric. Res. 6:813-816. 1916.

Rye Smut. (with M. N. Levine) Minn. Agric. Exp. Stn. Bull. 160. 1916.

Fruit and Vegetable Diseases and Their Control. (with A. G. Tolaas) Minn. Agric. Exp. Stn. Bull. 153. 1916.

Potato Diseases and Their Control. (with A. G. Tolaas) Minn. Agric. Exp. Stn. Bull. 158. 1916.

Diseases of fruit trees and plants. Fruitman Gardener 19:3-4. 1916.

Biologic forms of *Puccinia graminis* on wild grasses and cereals. (with F. J. Piemeisel) (Abstr.) Phytopathology 6:99-100. 1916.

The Prevention of Smuts. Minn. Agric. Ext. Div. Spec. Bull. 16. 1917.

Biologic forms of *Puccinia graminis* on cereals and grasses. (with F. J. Piemeisel) J. Agric. Res. 10:429-495. 1917.

Profit in treating fall-sown grains. Minn. Agric. Ext. Div., University Farm Press News, Vol. 8, No. 15. 1917.

A new strain of *Puccinia graminis*. (with F. J. Piemeisel) (Abstr.) Phytopathology 7:73. 1917.

Fruit tree cankers and their control. (with A. G. Newhall) Minn. State Entomol. Circ. 51. 1918.

The control of brown rot of plums and plum pocket. (with A. G. Tolaas) Minn. Hortic. 46:182-186. 1918.

The occurrence of *Puccinia graminis tritici-compacti* in the southern United States. (with

G. R. Hoerner) Phytopathology 8:141-149. 1918.

A third biologic form of *Puccinia graminis* on wheat. (with M. N. Levine) J. Agric. Res. 13:651-654. 1918.

Can biologic forms of stemrust on wheat change rapidly enough to interfere with breeding for rust resistance? (with John H. Parker and F. J. Piemeisel) J. Agric. Res. 14:111-124. 1918.

Plasticity of biologic forms of *Puccinia graminis*. (with F. J. Piemeisel and M. N. Levine) J. Agric. Res. 15:221-250. 1918.

The Black Stem Rust and the Barberry. U.S. Dept. Agric. Yearbook 1918:75-100. (Separate 796) 1919.

Destroy the Common Barberry. U.S.D.A. Farmers' Bull. 1058 (rev.). 1919.

Rust resistance in timothy. (with H. K. Hayes) J. Am. Soc. Agron. 11:67-70. 1919.

Controlling flax wilt by seed selection. (with H. K. Hayes, O. S. Aamodt, and J. G. Leach) J. Am. Soc. Agron. 11:291-298. 1919.

Effect of certain ecological factors on the morphology of the urediniospores of *Puccinia graminis*. (with M. N. Levine) J. Agric. Res. 16:43-77. 1919.

New biologic forms of *Puccinia graminis*. (with M. N. Levine and J. G. Leach) J. Agric. Res. 16:103-105. 1919.

Fruit tree cankers and their control. Minn. Hortic. 47:256-262. 1919.

Black rust of wheat. Spec. Circ. Spring Wheat Crop Improvement Assoc., Minneapolis, MN. 1920.

Puccinia graminis on native *Berberis canadensis*. (with L. J. Krakover) Phytopathology 10:305-306. 1920.

The regional occurrence of *Puccinia graminis* on barberry. (with R. S. Kirby and A. F. Thiel) (Abstr.) Phytopathology 11:39-40. 1921.

Resistance of barley to *Helminthosporium sativum* P. K. B. (with H. K. Hayes) Phytopathology 11:405-411. 1921.

Grain rust a menace in the Northwest. The Breeder's Gazette 79:992-993. 1921.

Black stem rust of wheat and how barberry spreads it. Farm, Stock Home 37:292-293. 1921.

Routing the black stem rust of wheat. Iowa Homestead 46:937. 1921.

Spring wheat's worst enemy. Furrow 26:5. (Deere and Webber Co., Minneapolis, MN.) 1921.

Wheat stem rust from the standpoint of plant breeding. (with H. K. Hayes) Proc. Annu. Meeting, 2nd. pp. 22-35. West. Can. Soc. Agron. 1921.

Review of *Diseases of Economic Plants* by F. L. Stevens and J. G. Hall. Bot. Gaz. 72:261-262. 1921.

The effect of fertilizers on the development of stem rust of wheat. (with O. S. Aamodt) (Abstr.) Phytopathology 12:31. 1922.

Observations on the spore content of the upper air. (with A. W. Henry, W. N. Christopher, and G. C. Curran) (Abstr.) Phytopathology 12:44. 1922.

Fruit and Vegetable Diseases. (with J. G. Leach and J. L. Seal) Minn. Agric. Exp. Stn. Bull. 199. 1922.

Prevent diseases of garden crops; It pays. Minn. Hortic. 50:215-217. 1922.

The Determination of Biologic Forms of *Puccinia graminis* on *Triticum* Spp. (with M. N. Levine) Minn. Agric. Exp. Stn. Tech. Bull. 8. 1922.

Fighting rust in Europe. Proc., Annu. Conf. for the Prevention of Grain Rust, 2nd. pp. 23-32. 1922.

Biologic specialization of *Puccinia graminis avenae*. (with M. N. Levine and D. L. Bailey) (Abstr.) Phytopathology 13:35. 1923.

Biologic specialization of *Puccinia graminis secalis*. (with M. N. Levine) (Abstr.) Phytopathology 13:35. 1923.

The species concept from the point of view of a plant pathologist. Am. J. Bot. 10:239-244. 1923.

Dusting Seed Grain to Prevent Smut. (with E. B. Lambert) Minn. Agric. Ext. Div. Spec.

Bull. 70. 1923.

Barberry eradication prevents black rust in western Europe. U.S.D.A. Circ. 269. 1923.

Health protection for wheat. Baking Technol. 2:54-57. 1923.

Spores in the upper air. (with A. W. Henry, G. C. Curran, and W. N. Christopher) J. Agric. Res. 24:599-606. 1923.

Destroy the Common Barberry. U.S.D.A. Farmers' Bull. 1058 (rev.). 1923.

Biologic forms of *Puccinia graminis* on varieties of *Avena* spp. (with M. N. Levine and D. L. Bailey) J. Agric. Res. 24:1013-1018. 1923.

European travels of a cereal pathologist. Trans. Wis. State Hortic. Soc. pp. 113-119. 1923.

Review of *Helminthosporium foot rot of wheat, with observations on the morphology of Helminthosporium and on the occurrence of saltation in the genus*, by F. L. Stevens. Bot. Gaz. 76:318-319. 1923.

Reaction of Barley Varieties to *Helminthosporium sativum*. (with H. K. Hayes, F. Griffee, and J. J. Christensen) Minn. Agric. Exp. Stn. Tech. Bull. 21. 1923.

The wheat rust problem in the United States. Proc. Pan-Pacific Sci. Congr. (Aust.). Vol. 1, pp. 88-96. 1923.

Methods of reducing losses from black stem rust of wheat. Proc. Pan-Pacific Sci. Congr. (Aust.). Vol. 1, pp. 132-136. 1923.

Some problems in plant quarantine. Proc. Pan-Pacific Sci. Congr. (Aust.). Vol. 1, pp. 163-170. 1923.

A Study of the Damping-Off Disease of Coniferous Seedlings. (with T. S. Hansen, W. H. Kenety, and G. H. Wiggin) Minn. Agric. Exp. Stn. Tech. Bull. 15. 1923.

Puccinia graminis on *Poa* spp. in the United States. (with M. N. Levine) (Abstr.) Phytopathology 14:39. 1924.

The effect of fertilizers on the development of stem rust of wheat. (with O. S. Aamodt) J. Agric. Res. 27:341-380. 1924.

Reactions of selfed lines of maize to *Ustilago zeae*. (with H. K. Hayes, F. Griffee, and J. J. Christensen) Phytopathology 14:268-280.

The present status of the cereal rust situation in the United States. Proc. Cereal Rust Conf., 2nd. pp. 9-26. Fed. Dept. Agric. and Res. Counc. Canada. 1924.

Cereal rust investigations in the United States. Proc. Cereal Rust Conf., 2nd. pp. 69-78. Fed. Dept. Agric. and Res. Counc. Canada. 1924.

Varietal resistance of spring wheats to *Tilletia levis*. (with E. B. Lambert and H. H. Flor) Minn. Studies in Plant Science, Studies in the Biological Sciences, No. 5. pp. 307-317. 1924.

Puccinia graminis poae Erikss. and Henn. in the United States. (with M. N. Levine) J. Agric. Res. 28:541-548. 1924.

Foot and root rots of wheat in Minnesota in 1924. (with J. J. Christensen) (Abstr.) Phytopathology 15:53. 1925.

The control of flax rust. (with A. W. Henry) (Abstr.) Phytopathology 15:53. 1925.

The control of loose smuts of wheat and barley, and barley stripe by Uspulun, Semesan, and Germisan. (with H. A. Rodenhiser) (Abstr.) Phytopathology 15:51. 1925.

Inheritance in wheat of resistance to black stem rust. (with H. K. Hayes and O. S. Aamodt) Phytopathology 15:371-387. 1925.

Raspberry mosaic. Minn. Hortic. 53:85-87. 1925.

Webster, a common wheat resistant to black stem rust. (with M. N. Levine and F. Griffee) Phytopathology 15:691-698. 1925.

Physiologic specialization of *Ustilago zeae* and *Puccinia sorghi* and their relation to corn improvement. (with J. J. Christensen) (Abstr.) Phytopathology 16:84. 1926.

The present status of the black stem rust situation. Northwest. Miller. Feb. 3, p. 476. 1926.

Relative susceptibility of spring-wheat varieties to stem rust. (with J. A. Clark and J. H. Martin) U.S. Dept. Agric. Circ. 365. 1926.

Preventing grain smut. Grain Dealers J. 56:225, 237. 1926.

Physiologic specialization of *Fusarium lini*, Bolley. (with W. C. Broadfoot) (Abstr.) Phytopathology 16:84-85. 1926.

Effect of sulfur dust on the development of black stem rust of wheat in a natural epidemic. (with E. B. Lambert) (Abstr.) Phytopathology 16:64-65. 1926.

Physiologic specialization and mutation in *Ustilago zeae*. (with J. J. Christensen) Phytopathology 16:979-999. 1926.

Treatments for seed grain. (with R. C. Rose and H. A. Rodenhiser) Leaflet issued by C. B. Lyon Co., St. Paul. 1926.

Stem Rust in Many Varieties Attacks Grain. U.S. Dept. Agric. Yearbook 1926:693-694. (Separate 961) 1926.

The common barberry and black stem rust. S. D. Educ. Assoc. J. 2:429-431. 1927.

Prevent smuts of small grains. (with H. A. Rodenhiser) Minn. Agric. Exp. Stn. Circ. 24. 1927.

Physiologic specialization in *Tilletia levis* and *Tilletia tritici*. (with H. A. Rodenhiser) Phytopathology 17:247-253. 1927.

Barberry Eradication Pays. (with L. W. Melander and D. G. Fletcher) Minn. State Dept. Agric. Bull. 55. 1927.

Flax resistant to wilt and sown early helps to cut losses. U.S. Dept. Agric. Yearbook. pp. 305-307. 1927. (Separate 1008, 1928)

Susceptibility of wheat varieties and hybrids to wheat scab in Minnesota. (with J. J. Christensen) (Abstr.) Phytopathology 17:40-41. 1927.

The control of apple rust. Minn. Hortic. 55:234-238. 1927.

The Common Barberry and Black Stem Rust. (with F. E. Kempton and L. D. Hutton) U.S. Dept. Agric. Farmers' Bull. 1544. 1927.

Heterothallism in *Ustilago zeae*. (with J. J. Christensen) Phytopathology 17:827-834. 1927.

Physiologic specialization in *Puccinia sorghi*. (with J. J. Christensen and H. E. Brewbaker) Phytopathology 18:345-354. 1928.

The relation of temperature during the growing season in the spring wheat area of the United States to the occurrence of stem rust epidemics. (with E. B. Lambert) Phytopathology 18:369-374. 1928.

The interdependence of the geneticist and the pathologist in wheat breeding, and their way of working together. In: Rep. Annu. Hard Spring Wheat Conf., 1st. N. D. Agric. College, Fargo. pp. 35-38. 1928.

Racial specialization in plant disease fungi. In: Lectures on Plant Pathology and Physiology in Relation to Man. (Mayo Foundation Lectures, 1926-1927). pp. 93-150. 1928.

Mutation in *Ustilago zeae*. (with J. J. Christensen and W. F. Hanna) (Abstr.) Phytopathology 19:106. 1929.

Physiologic specialization in plant pathogenic fungi. Leopoldina 4:263-289. 1929.

Susceptibility of Wheat Varieties and Hybrids to Fusarial Head Blight in Minnesota. (with J. J. Christensen and F. R. Immer) Minn. Agric. Exp. Stn. Tech. Bull. 59. 1929.

Sulphur dusting for the prevention of stem rust of wheat. (with E. B. Lambert) Phytopathology 19:631-643. 1929.

The value of physiologic-form surveys in the study of the epidemiology of stem rust. (with M. N. Levine and J. M. Wallace) Phytopathology 19:951-959. 1929.

Control of barley stripe. (with H. A. Rodenhiser) Minn. Agric. Ext. Div. Circ. 31. 1929.

Mutation and Hybridization in *Ustilago zeae*. Part I. Mutation. (with J. J. Christensen, C. J. Eide, and B. Peturson) Minn. Agric. Exp. Stn. Tech. Bull. 65. pp. 1-66. 1929.

Physiologic specialization in pathogenic fungi. Proc. Int. Congr. Plant Sci. Ithaca, N.Y. 1926. 2:1312-1330. 1929.

Field Studies on the Rust Resistance of Oat Varieties. (with M. N. Levine and T. R. Stanton) U.S. Dept. Agric. Tech. Bull. 143. 1930.

Barberry Eradication Pays. (with L. W. Melander and D. G. Fletcher) Minn. State Dept. Agric. Bull. 55 (rev.). 1930.

Black Stem Rust of Cereals Has More Than 60 Physiologic Forms. (with M. N. Levine) U.S. Dept. Agric. Yearbook 1930:137-140. (Separate 1135) 1930.

Flax Facts. (with A. C. Arny, H. A. Rodenhiser, et al.) Minn. Ext. Div. Spec. Bull. 128. Issued by Agric. Ext. Divs. of N. D., Mont., and S. D. Exp. Stns. and U.S. Dept. Agric. 1930.

Origin of physiologic forms of *Puccinia graminis* through hybridization and mutation. (with M. N. Levine and R. U. Cotter) Sci. Agric. 10:707-720. 1930.

Wheat protected from black stem rust by dusting with sulphur. (with L. H. Person) U.S. Dept. Agric. Yearbook 1930:547-548. 1930.

Racial specialization of parasitic fungi. Trans. Ky. Acad. Sci. 3:93-96. 1930.

Hybridization and mutation in *Puccinia graminis*. (with M. N. Levine and R. U. Cotter) (Abstr.) Phytopathology 20:113. 1930.

The Common Barberry and Black Stem Rust. (with D. G. Fletcher) U.S. Dept. Agric. Farmers' Bull. 1544 (rev.). 1930.

Die Bedeutung eines Pflanzenschutzgesetzes für die europäischen Länder. Mitt. DLG. 46:515-518, 540-543. 1931.

Review of *Handbuch der Pflanzenkrankheiten, Band II, Fünfte Auflage* by Paul Sorauer. 1928. Phytopathology 21:1093-1094. 1931.

Review of *The Plant Rusts (Uredinales)* by Joseph C. Arthur et al. Phytopathology 21:1205-1207. 1931.

Dissemination of cereal rusts. Proc. Int. Bot. Congr., 5th. Cambridge. 1930. pp. 411-413. 1931.

Distribution of physiologic forms of *Puccinia graminis tritici* in the United States and Mexico in relation to rust epidemiology. (with L. Hines, H. G. Ukkelberg, and W. Butler) (Abstr.) Phytopathology 22:25. 1932.

Physiologic forms of *Puccinia graminis* produced on barberries in nature. (with L. Hines, R. U. Cotter, and M. N. Levine) (Abstr.) Phytopathology 22:25. 1932.

Many are called but few are chosen. Gamma Alpha Record 22:73-74. 1932.

Spray program for the Northwest. (with A. G. Ruggles) Northwest. Druggist 40:35. 1932.

Problems in the genetics of phytopathogenic fungi. Proc. Int. Congr. Genet., 6th. 2:190-192. 1932.

Review of *Biologie der Pflanzenbewohnenden Parasitischen Pilze* by E. Fischer and E. Gäumann. (with C. R. Orton) Phytopathology 22:721-722. 1932.

Distribution of physiologic forms of *Puccinia graminis tritici* in relation to stem-rust epidemiology in 1932. (with L. Hines, H. G. Ukkelberg, and W. Butler) (Abstr.) Phytopathology 23:34. 1933.

The control of barley scab. Coop. Manager Farmer 22:34. 1933.

The constancy of cultural characters and pathogenicity in variant lines of *Ustilago zeae*. (with L. J. Tyler and G. E. Hafstad) Bull. Torrey Bot. Club 60:565-572. 1933.

Reaction of maize seedlings to *Gibberella saubinetii*. (with H. K. Hayes and I. J. Johnson) Phytopathology 23:905-911. 1933.

Fungi and bacteria on barley. (with J. J. Christensen) (Abstr.) Phytopathology 24:4-5. 1934.

The production of an apparently new variety of *Puccinia graminis* by hybridization on barberry. (with M. N. Levine and R. U. Cotter) (Abstr.) Phytopathology 24:13-14. 1934.

The pathogenicity and cytology of *Urocystis occulta*. (with M. B. Moore and R. C. Cassell) (Abstr.) Phytopathology 24:18. 1934.

Correlated inheritance of reaction to stem rust, leaf rust, bunt, and black chaff in spring-wheat crosses. (with H. K. Hayes, E. R. Ausemus, and R. H. Bamberg) J. Agric. Res. 48:59-66. 1934.

Uniform disease nurseries. (with M. N. Levine) In: Rep. Hard Spring Wheat Conference, 4th. Feb. 15-17. pp. 6-8. 1934.

Relation of barberry to the origin and persistence of physiologic forms of *Puccinia graminis*. (with M. N. Levine, R. U. Cotter, and L. Hines) J. Agric. Res. 48:953-969. 1934.

The cytology of *Urocystis occulta*. (with R. C. Cassell and M. B. Moore) Phytopathology

24:874-889. 1934.

Epidemiology of cereal rusts. Proc. Sci. Congr., 5th. Canada. 1933. 4:4177-3184. The University of Toronto Press. 1934.

Uniform rust nurseries indicate decreasing severity of stem rust. (with M. N. Levine) (Abstr.) Phytopathology 25:25. 1935.

Experiments on physiologic specialization and nature of variation in *Ustilago zeae*. (with L. J. Tyler, G. E. Hafstad, and E. G. Sharvelle) (Abstr.) Phytopathology 25:34. 1935.

Population trends of physiologic forms of *Puccinia graminis tritici*, 1930 to 1934. (with L. Hines, R. C. Cassell, and M. N. Levine) (Abstr.) Phytopathology 25:34. 1935.

The rusts of cereal crops. (with H. B. Humphrey, E. B. Mains, C. O. Johnston, H. C. Murphy, and W. M. Bever) U.S. Dept. Agric. Circ. 341. 1935.

Relation of *Fusarium* and *Helminthosporium* in barley seed to seedling blight and yield. (with J. J. Christensen) Phytopathology 25:309-327. 1935.

Progress and problems in breeding disease-resistant cereals for the spring wheat region of the United States. Proc. Zesde Int. Bot. Congr. Amsterdam. 1935. 2:49-51. 1935.

Physiologic specialization in cereal smuts. Proc. Zesde Int. Bot. Congr. Amsterdam. 1935. 2:187-188. 1935.

Die Bestimmung physiologischer Rassen pflanzenpathogener Pilze. (with M. N. Levine, J. J. Christensen, and K. Isenbeck) Nova Acta Leopoldina (Neue Folge) 3:281-336. 1935.

Stem rust in 1935. Mimeographed. 7 pp. Oct. 15, 1935.

A review of the aims, accomplishments and objectives of the barberry eradication program. Mimeographed. 20 pp. 1936.

Thatcher Wheat. (with H. K. Hayes, E. R. Ausemus, C. H. Bailey, H. K. Wilson, R. H. Bamberg, M. C. Markley, R. F. Crim, and M. N. Levine) Minn. Agric. Exp. Stn. Bull. 325. 1936.

The problem of specialization and variation in phytopathogenic fungi. Proc. Zesde Int. Bot. Congr. Amsterdam. 1935. vol. 1. pp. 30-35. 1936.

The problem of specialization and variation in phytopathogenic fungi. Genetica 18:372-389. 1936.

Biological problems in agriculture. (Address delivered at the 164th dinner of the New York Farmers, April 14.) Proc. New York Farmers, Season 1935-36. pp. 50-73. 1936.

The nature of resistance of cereals to rust. (with H. Hart) Rep. Int. Congr. Comp. Pathol., 3rd. Athens. 1936. vol. 1, pt. 2, pp. 253-266. 1936.

Variation in *Ustilago zeae*. (Abstr.) Science 85:58-59. 1937.

The promise of modern botany for man's welfare through plant protection. Sci. Monthly 44:117-130. 1937.

Case of respiratory allergy due to inhalation of grain smuts. (with F. W. Wittich) J. Allergy 8:189-193. 1937.

Value of research in agriculture. Minn. Alumni Weekly 37:197-198, 207. 1937.

The increase and importance of Race 56 of *Puccinia graminis tritici*. (with R. C. Cassell) (Abstr.) Phytopathology 28:20. 1938.

The epidemiology of stem rust of wheat in three successive contrasting years. (with W. T. Butler, R. U. Cotter, and J. J. Christensen) (Abstr.) Phytopathology 28:20. 1938.

Pathologische Probleme bei der Züchtung krankheitswiderstandsfähiger Weizen-und Gerstensorten in Sommerweizengebiet der Vereinigten Staaten von Nord-Amerika. (with J. J. Christensen and H. Becker) Zuchter 10:57-68. 1938.

Plant disease fungi constantly evolving new types. Science 88:438-439. 1938.

Observations on stem-rust epidemiology in Mexico. (with W. L. Popham and R. C. Cassell) (Abstr.) Phytopathology 29:22. 1939.

Observaciones del chahuixtle en Mexico durante el mes de Febrero de 1938. (with W. L. Popham and R. C. Cassell) Agricultura 2:3-5. 1939.

Studies of Inheritance in Crosses Between Bond, *Avena byzantina*, and Varieties of *A. sativa*. (with H. K. Hayes and M. B. Moore) Minn. Agric. Exp. Stn. Tech. Bull. 137. 1939.

Parasitic strains of plant disease fungi in relation to quarantines and eradication programs. (Abstr.) Proc. Annu. Meet., 15th. Central Plant Board. St. Paul, MN. Mar. 21-22. pp. 51-52. 1939.

Natural resistance, including immunity: Plant kingdom. (Abstr.) Abstracts of Communications. Third Int. Congr. Microbiol. New York. 1939. p. 217. 1939.

Stem rust in 1938. (with L. M. Hamilton) U.S. Dept. Agric. Plant Dis. Rep. Suppl. 117:69-83. 1939.

Population trends of physiologic races of *Puccinia graminis tritici* in the United States from 1930 to 1939. (with R. C. Cassell and W. Q. Loegering) (Abstr.) Phytopathology 30:22-23. 1940.

Observations on stem rust epidemiology in Mexico. (with W. L. Popham and R. C. Cassell) Am. J. Bot. 27:90-99. 1940.

The need for research on the genetics of pathogenic organisms. In: The Genetics of Pathogenic Organisms. Publ. of Am. Assoc. Adv. Sci. No. 12. pp. 9-18. Science Press Printing Co., Lancaster, PA. 1940.

Summary. In: The Genetics of Pathogenic Organisms. Publ. of Am. Assoc. Adv. Sci. No. 12. pp. 83-90. 1940.

Report on Tingo Maria and Satipo regions. Andean Air Mail and Peruvian Times 1:27-28. 1941.

Physiologic races of *Puccinia graminis* in the United States in 1939. (with W. Q. Loegering) Multigraphed. 14 pp. U.S. Dept. Agric., Bur. Entomol. Plant Quarantine E-522. 1941.

Biotypes within *Puccinia graminis tritici*, Race 15. (with W. Q. Loegering) (Abstr.) Phytopathology 32:12-13. 1942.

Recent changes in prevalence of physiologic races of *Puccinia graminis tritici* in the United States. (with W. Q. Loegering) (Abstr.) Phytopathology 32:17. 1942.

Physiologic races of *Puccinia graminis* in the United States in 1940. (with W. Q. Loegering and R. U. Cotter). Multigraphed. 20 pp. U.S. Dept. Agric., Bur. Entomol. Plant Quarantine E-522-A. 1942.

Physiologic races of *Puccinia graminis* in the United States in 1941. (with W. Q. Loegering) Multigraphed. 12 pp. U.S. Dept. Agric., Bur. Entomol. Plant Quarantine E-522-B. 1942.

The field of extramural aerobiology. In: Aerobiology. Publ. of Am. Assoc. Adv. Sci. No. 17. pp. 1-17. 1942.

Genetic variation in plant pathogens and its practical importance. Proc. Inst. Med. Chicago 14:250-260. 1942.

An unusually virulent race of wheat stem rust, No. 189. (with G. Garcia-Rada, J. Vallega, and W. Q. Loegering) Phytopathology 32:720-726. 1942.

Genetic factors for mutability and mutant characters in *Ustilago zeae*. (with M. F. Kernkamp, T. H. King, and W. J. Martin) Am. J. Bot. 30:37-48. 1943.

Regional spread of wheat stem rust from barberry-infested areas of the Virginias in 1942. (with R. U. Cotter and W. Q. Loegering) (Abstr.) Phytopathology 33:12. 1943.

Physiologic races of *Puccinia graminis* in the United States in 1942. (with W. Q. Loegering) Multigraphed. 9 pp. U.S. Dept. Agric., Bureau of Entomol. and Plant Quarantine E-522-C. 1943.

Population trends of physiologic races of *Puccinia graminis tritici* in the United States for the period 1930 to 1941. (with W. Q. Loegering, R. C. Cassell, and L. Hines) Phytopathology 33:884-898. 1943.

The inheritance of a white mutant character in *Ustilago zeae*. (with M. F. Kernkamp, W. J. Martin, and T. H. King) Phytopathology 33:943-949. 1943.

Races of *Puccinia graminis avenae* in the United States. (with M. N. Levine and W. Q. Loegering) (Abstr.) Phytopathology 33:1118. 1943.

Plant diseases are shifty enemies. Minn. Farm Home Sci. 2(1):8-9, 12. 1944.

Newthatch wheat—What it is and what it took to make it. (with E. R. Ausemus) Minn. Farm Home Sci. 2(1):13. 1944.

Newthatch Wheat. (with E. R. Ausemus, E. W. Hanson, W. F. Geddes, and P. P. Merritt) Minn. Agric. Exp. Stn. Tech. Bull. 166. 1944.

Relation of physiologic races of *Puccinia graminis tritici* to wheat improvement in Southern Mexico. (with J. G. Harrar and W. Q. Loegering) (Abstr.) Phytopathology 34:1002. 1944.

Identification of Physiologic Races of *Puccinia graminis tritici*. (with M. N. Levine and W. Q. Loegering) U.S. Dept. Agric.: Agric. Res. Admin., Bur. Entomol. Plant Quarantine E-617. 1944. [Copies also printed with illustrations and color plates added, by the Conference for the Prevention of Grain Rust, Minneapolis, Minn.]

The potential importance of race 8 of *Puccinia graminis avenae* in the United States. (with W. Q. Loegering) Phytopathology 34:421-425. 1944.

Physiologic races of *Puccinia graminis* in the United States in 1943. (with W. Q. Loegering) Processed. 10 pp. U.S. Dept. Agric.: Bur. Entomol. Plant Quarantine; Bur. Plant Ind., Soils, Agric. Eng.; and Minnesota Agric. Exp. Stn. 1945.

Plant pathology in Mexico. (with J. G. Harrar) In: Plants and Plant Science in Latin America. F. Verdoorn, ed. pp. 52-55. The Chronica Botanica Co., Waltham, MA. 1945.

The field reactions of certain oats hybrids and varieties to stem rust. (with M. B. Moore and H. K. Hayes) Phytopathology 35:549-551. 1945.

Arthur Henry Reginald Buller. (with W. F. Hanna and C. W. Lowe) Phytopathology 35:577-584. 1945.

Herbert Hice Whetzel. (with M. F. Barrus) Phytopathology 35:659-670. 1945.

Aerobiology in relation to plant disease. (with C. M. Christensen) Bot. Rev. 12:205-253. 1946.

Flag smut of wheat in Mexico. (with N. E. Borlaug and J. G. Harrar) Phytopathology 36:479. 1946.

Recent changes in the stem-rust situation in Mexico. (with J. G. Harrar and W. Q. Loegering) (Abstr.) Phytopathology 36:400. 1946.

Adaptation of monosporidial lines of *Ustilago zeae* to arsenic. (with F. V. Stevenson and C. T. Wilson) (Abstr.) Phytopathology 36:411. 1946.

Plant pathologist's merry-go-round. J. Hered. 37:259-265. 1946.

Fungi—Friends and foes. In: The Scientists Speak. pp. 196-199. Boni and Gaer, New York, 1947. (One of a series of radio talks broadcast by American scientists, on the New York Philharmonic-Symphony radio program, sponsored by United States Rubber Company, 1946.)

International problems in plant disease control. Proc. Am. Philos. Soc. 91:95-111. 1947.

The nature and importance of physiologic specialization in phytopathogenic fungi. Science 105:627-632. 1947.

Plant diseases are shifty enemies. Am. Sci. 35:321-350. 1947.

Plant diseases are shifty enemies. In: Science in Progress, Fifth Series. G. A. Baitsell, ed. pp. 235-279. Yale University Press, New Haven. 1947.

Review of *Pflanzliche Infektionslehre* by Ernst O. Gäumann. Arch. Biochem. 13:304-306. May 1947. Also Phytopathology 37:684-685. Sept. 1947.

Variation in phytopathogenic fungi. (with C. M. Christensen and J. J. Christensen) Annu. Rev. Microbiol. 1:61-84. 1947.

The relation of barberries to physiologic races of *Puccinia graminis* in the United States in recent years. (with W. Q. Loegering) (Abstr.) Phytopathology 38:24. 1948.

Reaction of wheat varieties in the seedling stage to races of *Puccinia graminis tritici* in greenhouse tests. (with W. Q. Loegering) Mimeographed. 9 pp. 1948.

Physiologic races of *Puccinia graminis* in the United States in 1944. (with W. Q. Loegering) Processed. 8 pp. U.S. Dept. Agric.: Agric. Res. Admin.; Bur. Entomol. Plant Quarantine; Bur. Plant Ind., Soils, Agric. Eng.; and Minn. Agric. Exp. Stn. 1948.

Variation induced by uranium nitrate in corn smut and the cultivated mushroom. (with J. M. Daly, M. L. Gattani, and I. Wahl) Science 108:554-555. 1948.

Review of *Botanik der Gegenwart und Vorzeit in culturhistorische Entwickelung* by Karl F. W. Jessen. Sci. Monthly 67:445. 1948.

Science and human subsistence. Mimeographed. Reprint with slight modification from New England Homestead (Springfield, MA). Vol. 121, Dec. 11, 1948.

Recent changes in prevalence of physiologic races of *Puccinia graminis tritici* in South-Central Mexico. (with J. G. Harrar, W. Q. Loegering, and N. E. Borlaug) (Abstr.) Phytopathology 39:23. 1949.

Science in the service of agriculture. Sci. Monthly 68:75-83. 1949.

Stem rust and barberry eradication. (Abstr.) Proc. Annu. Meet., 25th. Central Plant Board, Milwaukee, WI. March 22-23, 1949. pp. 52-54. 1949.

Science and its sphere of influence. Gamma Alpha Record 39:62-69. 1949.

Review of *The Life of Science: Essays in the History of Civilization* by George Sarton. Am. Hist. Rev. 54:851-853. 1949.

Soil nutrients as affecting susceptibility to plant disease. (with J. M. Daly) (Abstr.) Abstracts, Annu. Meet. Am. Soc. Agron. and Soil Sci. Soc. Am., 41st. Milwaukee, Wis. Oct. 24-28, 1949. pp. 100-101. 1949.

Science—Its sphere of influence. Research 3:101-105. 1950.

Science and national authoritarianism. Proc. Am. Philos. Soc. 94:114-120. 1950.

The present status of breeding rust resistant oats at the Minnesota Station. (with W. R. Kehr, H. K. Hayes, and M. B. Moore) Agron. J. 42:356-359. 1950.

Razas fisiológicas de *Puccinia graminis tritici* en México. (with W. Q. Loegering, J. G. Harrar, and N. E. Borlaug) Oficina de Estudios Especiales, Secretaria de Agricultura y Ganaderia, Mexico, D. F. Folleto Tecnico No. 3. 1950.

Varietal resistance of spring wheats to fusarial head blight. (with E. W. Hanson and E. R. Ausemus) Phytopathology 40:902-914. 1950.

New races of stem rust break out. Minn. Farm and Home Sci. 8(1):7, 15. 1950.

Physiologic races of *Puccinia graminis* in the United States in 1945-49. (with W. Q. Loegering) Processed. 19 pp. U.S. Dept. Agric.: Bur. Entomol. Plant Quarantine; Bur. Plant Ind., Soils, Agric. Eng.; and Minn. Agric. Exp. Stn. 1950.

The control of plant diseases. Proc. U.N. Sci. Conf. on Conservation and Utilization of Resources. Lake Success, N.Y. Aug. 17-Sept. 6, 1949. 6:319-324. 1951.

Increase in prevalence of *Puccinia graminis tritici* Race 15B and *P. graminis avenae* Race 7 in 1950. (with W. Q. Loegering) (Abstr.) Phytopathology 41:33. 1951.

Science "can serve significantly" through UNESCO to stimulate world cooperation, promote peace. UNESCO News 5:7-9. 1951.

Science and international understanding. School Sci. Math. 113:11-18. 1951.

Science and human affairs. Science 113:137-142. 1951.

Wheat Leaf Rust Studies in St. Paul, Minnesota. (with M. N. Levine and E. R. Ausemus) Plant Dis. Rep. Suppl. 199. 1951.

Physiologic races of *Puccinia graminis* in the United States in 1950. (with W. Q. Loegering) Processed. 16 pp. U.S. Dept. Agric.: Bur. Entomol. Plant Quarantine; Bur. Plant Ind., Soils, Agric. Eng.; and Minn. Agric. Exp. Stn. 1951.

Contributions of science to international understanding. Chicago Schools J. 33:90-92. 1952.

The mutagenic action of dilute colloidal polonium on fungi. (with J. B. Rowell and E. E. Butler) (Abstr. 6) U.S. Atomic Energy Comm. Nuclear Sci. Abstr. 4:138. 1952.

Free amino-acids and carbohydrates synthesized by *Ustilago zeae*. (with J. B. Rowell and J. E. DeVay) (Abstr.) Phytopathology 42:6. 1952.

The natural science program in UNESCO. AIBS Bull. 2:12-15. 1952.

Introductory statement. Congr. Int. Patologia Comparada, 6th. Madrid. May 4-11, 1952. Vol. I. Ponencias. pp. 3-4. 1952.

Characters of fungi important in comparative pathology. Congr. Int. Patologia Comparada, 6th. Madrid. May 4-11, l952. Vol. I. Ponencias. pp. 5-14. 1952.

The behavior of mutable black lines of *Ustilago zeae*. (with Shih I. Lu and J. B. Rowell) (Abstr.) Phytopathology 42:480. 1952.

Review of *Pflanzliche Infektionslehre* by E. Gäumann. (with H. Hart) Arch. Biochem. Biophys. 41:486-487. 1952.

The rust problem today. (Abstr.) In: Grain Rust Review of 1953. N. D. Agric. Exp. Stn. Bimonthly Bull. 15:196-199. 1953.

Physiologic races of *Puccinia graminis* in the United States in 1951. (with D. W. Stewart) Processed. 10 pp. U.S. Dept. Agric.: Agric. Res. Admin.; Bur. Entomol. Plant Quarantine; Bur. Plant Ind., Soils, Agric. Eng.; and Minn. Agric. Exp. Stn. 1953.

The present status of the stem rust problem. (with H. Hart and D. M. Stewart) In: Report of the International Wheat-Stem-Rust Conf. held at Fort Gary Hotel, Winnipeg, Manitoba, Canada, Jan. 5-7, 1953. pp. 2-5. Plant Ind. Stn., Beltsville, MD. 277CC. 1953.

Biotypes within races of stem rust. (with H. Hart, R. Postigo, and S. Goto) In: Rep. Int. Wheat-Stem-Rust Conf. held at Fort Gary Hotel, Winnipeg, Manitoba, Canada, Jan. 5-7, 1953. pp. 7-9. Plant Ind. Stn., Beltsville, MD. 277CC. 1953.

The pathogenicity of Races 11, 49, and 139 of *Puccinia graminis tritici* to Kenya wheat varieties and their derivatives. (with E. B. Hayden and D. H. Smith). In: Rep. Int. Wheat-Stem-Rust Conf. held at Fort Gary Hotel, Winnipeg, Manitoba, Jan. 5-7, 1953. p. 13. Plant Ind. Stn., Beltsville, MD. 277CC. 1953.

Variability in rust reaction. (with H. Hart). In: Rep. Int. Wheat-Stem-Rust Conf. held at Fort Gary Hotel, Winnipeg, Manitoba, Jan. 5-7, 1953. p. 19. Plant Ind. Stn., Beltsville, MD. 277CC. 1953.

The natural sciences program in UNESCO. Natl. Acad. Sci., Nat. Res. Counc., News Report 3:1-4. Jan.-Feb. 1953.

Physiologic races of *Puccinia graminis* in the United States in 1952. (with D. M. Stewart) Processed. 11 pp. U.S. Dept. Agric.: Bur. Entomol. Plant Quarantine; Bur. Plant Ind., Soils, Agric. Eng.; and Minn. Agric. Exp. Stn. 1953.

Problems in variability in fungi. (with J. J. Christensen) In: Plant Diseases. Yearbook of Agriculture 1953. pp. 35-62. U.S. Dept. Agric., Washington, DC. 1953. (Separate 2408, 1954)

The 1953 stem rust situation and the status of barberry eradication. (Abstr.) Exten. Conf., Nicollet Hotel, Minneapolis, Minn. Nov. 16-17, 1953. Northwest Crop Improvement Assoc., Minneapolis, Minn. 1953.

Amino-acids synthesized by lines of *Ustilago zeae* of different sex. (with J. E. DeVay) (Abstr.) Phytopathology 43:470. 1953.

The origin and extent of a regional spore shower of wheat stem rust. (with D. G. Fletcher, E. B. Lambert, and K. M. Nagler) (Abstr.) Phytopathology 43:471. 1953.

The pathogenicity of races 11, 49, 125 and 139 of *Puccinia graminis tritici* to Kenya wheat varieties and their derivatives. (with E. B. Hayden and D. H. Smith) (Abstr.) Phytopathology 43:474. 1953.

Radioactivity aids study of plant diseases. Minn. Farm Home Sci. 11(2):13. 1954.

Plant pathology through the years. Minn. Farm Home Sci. 11(2):20. 1954.

Physiologic races of *Puccinia graminis* in the United States in 1953. (with D. M. Stewart) Processed. 9 pp. U.S. Dept. Agric.: Agric. Res. Serv.; Plant Pest Control Branch; Field Crops Res. Branch; and Minn. Agric. Exp. Stn. 1954.

Recent studies of wheat stem rust in relation to breeding resistant varieties. Phytopathology 44:346-351. 1954.

Rust, a persistent threat to wheat. Agric. Food Chem. 2:14. 1954.

Edward Monroe Freeman, pioneer plant pathologist. (with J. J. Christensen) Science 120:285-286. 1954.

People, pathogens, and progress in international disease control. Phytopathology 44:421. 1954.

Progress and problems in plant pathology. Ann. Appl. Biol. 42:22-33. 1955.

Physiologic races of *Puccinia graminis* in the United States in 1954. (with D. M. Stewart, E. B. Hayden, and B. J. Roberts) Processed. 14 pp. U.S. Dept. Agric.: Agric. Res. Serv.; Plant Pest Control Branch; Field Crops Res. Branch; and Minn. Agric. Exp. Stn. 1955.

The new view of man in his biological environment. Centen. Rev. Arts Sci. 1:26-49. 1957.

Problems in preventing plant disease epidemics. Am. J. Bot. 44:259-267. 1957.

La lucha contra los males de los plantas. (with C. J. Eide) Mecanica Popular 20:81-84, 157-159. 1957.

Disease-free crops. (with C. J. Eide) Popular Mechanics 107:165-168, 252, 254. 1957.

Principles of Plant Pathology. (with J. G. Harrar) Ronald Press, New York. 581 pp. 1957.

B. M. Duggar, pioneer in precise plant research. Science 126:690. 1957.

The problem of classifying biotypes of *Puccinia graminis* var. *tritici* and other cereal rusts. (with D. M. Stewart and E. B. Hayden) (Abstr.) In: Rep. Int. Wheat Rust Conf., 3rd. Mexico City. Mar 18-24, 1956. pp. 97-99. Plant Ind. Stn., Beltsville, MD. 447CC. 1957.

Races of wheat stem rust in the United States and Mexico in 1955. (with D. M. Stewart and R. U. Cotter) (Abstr.) In: Rep. Int. Wheat Rust Conf., 3rd. Mexico City. Mar 18-24, 1956. pp. 99-100. Plant Ind. Stn., Beltsville, MD. 447CC. 1957.

Physiologic races of *Puccinia graminis tritici* collected from wheat near barberry bushes. (with W. Q. Loegering, D. M. Stewart, W. M. Watson, and C. W. Roane) (Abstr.) In: Rep. Int. Wheat Rust Conf., 3rd. Mexico City. Mar 18-24, 1956. pp. 112-113. Plant Ind. Stn., Beltsville, MD. 447CC. 1957.

Fortschritte und Probleme bei der Schaffung krankheitsresistenter Kulturpflanzen. Gesunde Pflanz. 9:225-235. 1957.

Fortschritte und Probleme bei der Schaffung krankheitsresistenter Kulturpflanzen. Umschau 1957:737-740. 1957.

The role of plant pathology in the scientific and social development of the world. AIBS Bull. 8:15-18. 1958.

Internationalism in plant pathology. Tijdschr. Planteziekten 64:354-356. 1958.

Race 15B of wheat stem rust—What it is and what it means. (with H. A. Rodenhiser) Adv. Agron. 10:143-165. 1958.

The role of plant pathology in the scientific and social development of the world. In: Plant Pathology, Problems and Progress 1908–1958. Univ. Wis. Press, Madison. pp. 3-13. 1959.

Progress and problems in the development of disease-resistant varieties of crop plants. Proc. Int. Congr. Crop Prot., 4th. Hamburg. 1957. vol. 1, pp. 5-14. 1959.

The need for international cooperation on cereal rusts. In: Omagiu lui Traian Savulescu. pp. 745-753. Academia Republicii Populare Romine, Bucharest. 1959.

Trends and needs in agricultural education and research. Proc. Am. Assoc. Land-Grant Colleges and State Universities. Washington, DC. Nov. 10-13, 1958. pp. 61-75. 1959.

Survival of physiologic races of *Puccinia graminis* var. *tritici* on wheat near barberry bushes. (with C. W. Roane, W. Q. Loegering, D. M. Stewart, and W. M. Watson) Phytopathology 50:40-44. 1960.

The problem of breeding resistant varieties. (with J. J. Christensen) In: Plant Pathology, Vol. III, The Diseased Population, Epidemics and Control. J. G. Horsfall, ed. pp. 567-624. Academic Press Inc., New York. 1960.

La obligación de la fitopatología en el problema de la alimentación humana. Memoria Congr. Natl. Entomol. Fitopatol., 2nd. Escuela Nacional de Agricultura, Chapingo, Mexico. 1960. pp. 479-501. 1960.

Importancia del estudio de la historia natural y su ensenanza. Rev. Soc. Mex. Hist. Nat. 21:485-508. 1960.

Botany and human affairs. In: Recent Advances in Botany. pp. 3-9. University of Toronto Press. 1961.

Identificazione di alcuni biotipi della razza fisiologica 21 di *Puccinia graminis* var. *tritici* in Italia nel 1960. (with A. L. Madaluni, R. Basile, and C. Sibilia) Boll. Stn. Patol. Veg. Rome. 19(Ser. 3):103-112. 1961.

Identification of Physiologic Races of *Puccinia graminis* var. *tritici*. (with D. M. Stewart and W. Q. Loegering). U.S. Dept. Agric., Agric. Res. Service E617 (rev.). 53 pp. 1962.

Can there be food for all? Proc. Fed. Sci. Congr., 1st. Salisbury, Southern Rhodesia. May 18-22, 1960. pp. 11-16. 1962.

Contribution of science to the understanding and control of plant diseases. Proc. U.N. Conf. on the Application of Science and Technology for the Benefit of the Less

Developed Areas. Geneva. 1963. vol. 3, pp. 152-158. 1963.

The role of plant pathology in the scientific and social development of the world. In: Selected Botanical Papers. Irving W. Knobloch, ed. pp. 132-138. Prentice-Hall, Inc., Englewood Cliffs, NJ. 1963.

Will the fight against wheat rust ever end? Z. Planzenkr. Pflanzenschutz 71:67-73. 1964.

Opportunity and obligation in plant pathology. Annu. Rev. Phytopathol. 2:1-12. 1964.

Science, Sense and Society. The Cosmos Club, Washington, DC. 22 pp. May 13, 1964.

Applied microbiology in the future of mankind. In: Proc. Global Impacts of Applied Microbiology. Stockholm, Sweden. July 29-Aug. 3, 1963. pp. 9-18. John Wiley & Sons Inc., New York. 1964.

Pest, pathogen, and weed control for increased food production. Proc. Natl. Acad. Sci. August, 1966.